*f*P

ALSO BY LEE ALAN DUGATKIN

Cheating Monkeys and Citizen Bees
(Free Press, 1999)

Game Theory and Animal Behavior
(Edited with Hudson Kern Reeve)

Cooperation Among Animals

The Imitation Factor

EVOLUTION BEYOND THE GENE

Lee Alan Dugatkin

THE FREE PRESS
New York London Toronto Sydney Singapore

THE FREE PRESS
A Division of Simon & Schuster, Inc.
1230 Avenue of the Americas
New York, NY 10020

Book design by Ellen R. Sasahara

Manufactured in the United States of America

1 3 5 7 9 10 8 6 4 2

Library of Congress Cataloging-in-Publication Data

Dugatkin, Lee Alan
The imitation factor : evolution beyond the gene / Lee Alan Dugatkin.
p. cm.
Includes bibliographical references (p.).
1. Animal behavior. I. Title.

QL751 D7465 2000
591.5—dc21 00-059286

ISBN 0-684-86453-3

To my mother, Marilyn,
who's a real peach

Contents

CONTENTS

Preface

WE DESPERATELY WANT TO THINK OF OURSELVES as somehow distinct from other life forms on our planet. We used to claim that we humans were unique in that we were the only species to make tools; now most of us have seen the nature programs on television featuring hammer-wielding monkeys and realize that this claim simply isn't true. Currently there is the sense that we are unique because "culture" is found only in humans. So where did culture come from? As we shall see, culture is not humanity's gift to the universe. It was invented long before we *Homo sapiens* arrived, and its power is evident in everything ranging from fish to nonhuman primates, and in virtually every behavioral context imaginable. It is not our gift, but it is our duty to understand it, to look into the heart of our social existence as individuals just as we have looked into the heart of our genetic existence as strands of DNA.

We have learned that the replication of DNA is a fundamental principle of the continuation of life on earth. But we have not recognized the fundamental importance of another natural copying mechanism: the individual. Although individuals are not replicators in the traditional sense, the fact that we are veritable copy machines in the wild is the key to understanding the development of human culture and to understanding an unrecognized pervasive force of nature: the imitation factor.

As is typical of any book that attempts to provide some kind of answer to a particularly large question, I am grateful for my interactions with hundreds of colleagues over the years and sincerely appreciate their work, thoughts, and encouragement.

I thank my editor at The Free Press, Stephen Morrow, for all his help. Stephen has been critical in shaping this book at every stage along the way, and I consider myself very lucky to have had him as my editor on this project.

Were it not for the advice and encouragement of my literary agent, Susan Rabiner, this work would never have come to be. Susan went through dozens of versions of early proposals for this book and was invaluable in helping me to see the forest through the trees.

My wife, Dana, has read every word in this book more times than anyone ever should. Dana is more than a proofreader, though. Her thoughts along the way have made this book much more accessible to a broad audience. In every way, she is an incredible helpmate. My five-year-old boy, Aaron, also proofread parts of this book (honest), but it is the sparkle in his eyes that I'm really grateful for.

Last, as always, I thank Dr. Jerram L. Brown for having confidence in my abilities early on. Fourteen years ago, Jerry believed in my talents, when in all honesty, there really was no reason to do so.

1

The Cultured Animal

> Imitation is natural to man from childhood, one of his advantages over the lower animals being this, that he is the most imitative creature in the world and learns at first by imitation.
>
> ARISTOTLE

ARISTOTLE RAISED ALL THE BEST QUESTIONS scientists are raising in laboratories today. That might be an overstatement, but if it is, it isn't a grand overstatement; it holds true for the basic questions that we address in my lab, as well as in the labs of many of my colleagues. But Aristotle was wrong, at least for the most part, about the ability of nonhuman animals to imitate. It was a big oversight. Nearly two and a half thousand years later, we discovered nature's blueprints—genes—but still remained almost blind to nature's way of transmitting information across the generations outside the genome. We tend to think we are the only animals able to do the trick of passing down the wisdom of our forebears. That trick is known as culture. Surprisingly, even guppies can do it.

Looking at infants and children, you can see that Aristotle is correct that humans learn first by imitation. A vast amount of preschool education comes from mimicking the behavior of others, usually parents. This is so obvious it can be scary. Looking at your child gives something of a mirror image of your own actions—and that isn't always a pleasant experience. No one doubts the importance of imitation in humans. But Aristotle did not really recognize imitation as the root of cultural transmission, nor did he see that many animals, not only ourselves, have been able to transmit culture in just this way. Indeed, no one at all has recognized this until very recently. But it is now a scientific fact that illuminates the mysterious origin of culture itself—and how evolution proceeds beyond the gene.

GUPPY CULTURE

I spend a lot of time watching guppies. To be honest, I spend more time with guppies than any sane person should, unless that person happens to be a behavioral ecologist—someone who studies the evolution of social behavior—and even then it is a close call. Watch guppies long enough, and you see that sex is what life in a guppy neighborhood (be it a tank or stream) revolves around. Males want to mate with any willing female, and spend a great deal of their time pursuing this desire. Females, on the other hand, spend the majority of their time avoiding the constant sexual harassment to which they are subjected. What determines who mates with whom in guppy society is a fascinating concoction of genetic drives and imitation. Females have a genetic predisposition to mate with colorful males, but layered on top of this is a strong inclination to imitate the choice of mates that other females make.

All else being equal, female guppies mate with colorful males and in essence obey their genetic code. But, hold everything constant, and females also imitate each other's choice of mating partner—and that alone is a rather remarkable finding. Female guppies, with brains no bigger than a pinhead, copy each other's choice of mates. Yet the story of copying and mate choice hardly ends there. One can move on to ask how genes and culture interact to shape guppy mate choice. That is, if we consider copying and imitation (and teaching) to be forces underlying the cultural transmission of behavior, how do these forces interact with genes coding for behavior?

We might think guppies alone should not prove the likes of Aristotle wrong. However, the role of culture in animal life has been observed in many many species, and its effects are clearly quite powerful. Various forms of culture influence mate choice in everything from bugs, fish, and birds to deer and primates, including, of course, humans. In many of these instances, culture and genetics interact in unexpected and bizarre ways. Furthermore, culture is in no way restricted to the subject of mate choice; it invades virtually every aspect of animal behavior. Culture, or more specifically cultural transmission of information through imitation, was a force long before humans came on the scene and continues to be an important factor in what many would consider to be very simple animals. And imitation is a unique factor because the actions of a few individuals, if copied, can have long-term evolutionary consequences at the population level. The quirky behavior of just a few individuals can live on across the generations and across thousands, potentially even millions, of years.

That culture plays such a strong role in animal choice is a controversial claim and an even more consequential fact. Consider that cultural transmission of information can work with lightning speed (when compared with the timescale on which genetic change takes place). The action of even a single individual, if it is copied by many, can snowball through a population with evolutionary reverberation. Moreover, new theory in biology can predict when cultural or genetic factors will predominantly drive behavior. That we can now balance hereditary forces against cultural ones is an astonishing achievement of modern science. But we need to step back and define what culture really is and answer some related questions before we consider the larger implications of this work.

Evolutionary biologists are generally in agreement that the process of natural selection is the primary (but not exclusive) agent driving evolutionary change. There are certainly some prominent critics of this view, but most biologists would not quibble with this statement, especially if one were talking about the evolution of social behavior. To understand virtually any aspect of life on earth or elsewhere, scientists will tell you, we must understand how the process of natural selection operates.

NATURAL SELECTION AND GENES

Given the monumental impact of his work in both the social and natural sciences, it is often surprising to learn that Charles Robert Darwin's ideas with respect to natural selection are remarkably straightforward. Consider any characteristic of an organism—height, weight, visual acuity, or anything else. If variations in this characteristic exist, such that, for example, there are differences in height among individuals in a population, and if there exists *a means by which individuals of different height produce offspring that resemble themselves,* then any variant that outreproduces others will spread through time.

If taller individuals, for whatever reason, have more offspring, over time we expect to see the average height of individuals in that population increase. If individuals who are slightly shorter produce more offspring, natural selection will, over time, methodically work toward producing a population of shorter individuals. This argument holds true even if there is a very slight edge in terms of the number of offspring that are raised. Through evolutionary time, small differences can accumulate into large changes. For example, if slightly taller individuals produce an average of 2.01 offspring per generation, with average individuals producing 2.00, then over time, natural selection will favor slightly taller individuals, and the population we are examining will be, on average, a bit taller than it used to be. It might take a while, but it will happen.

Darwin's theory was generic; it applied to any trait that met a few conditions he outlined in *The Origin of Species.* Behavioral biologists have long argued, as did Darwin himself, that the theory of natural selection is applicable not only to how an organism looks (anatomy, morphology, physiology), but to how it behaves as well. If a number of different behavioral options exist and *if there is some means for these behavioral options to be transmitted across generations,* then any behavior that has a slight advantage in terms of its effect on individual reproduction will increase in frequency.

Imagine that we have two different types of lions: some hunt prey by hiding in the brush, and others stand out in the open for all to admire their leonine majesty. Now suppose that each type produces offspring that use the same hunting strategy as their parents. If our ambushers do better than their nonambushing counterparts, ambush hunting behavior will be favored by natural selection and will, over time, increase in frequency. It is the same story with height; just replace height with ambush strategy.

It certainly seems reasonable to suggest that traits like height and hunting strategy are transmitted from one generation to the next by genes. Everyone knows that children look like their biological parents. Even if you have not taken a class in biology since high school, you know that this is so because parents pass their genes on to their offspring. But Darwin didn't know about genes when he published *The Origin of Species* in 1859. No one did. Until the turn of the twentieth century, the concept of genetic inheritance as we now understand it was not yet known by the vast majority of scientists. Nevertheless, once we found out the mechanisms of genetic inheritance, it made sense, at first glance, that these mechanisms would apply to all the traits Darwin thought were inherited.

Gregor Mendel, surely the most famous Austrian monk of all time, discovered the essential idea of the gene. He completed his simple and elegant experiments on the heredity of height in pea plants in 1865. Mendel's work was the start of the age of genetics, but despite the fact that Mendel did most of his work during the 1850s and 1860s, his findings were not disseminated until 1900. Darwin's grand theory predated the scientific community's understanding of what would be called the gene by forty years. In fact, the word *gene* was not used as a technical term for the unit of inheritance until 1909.

The story about genes and Darwin is strange. Not only was his theory of natural selection correct without precise information about genes; it was correct even though Darwin was just plain wrong about many aspects of inheritance. Darwin's ideas about how traits are passed down across generations focused on "gemmules." He believed that different parts of the body cast off many individual tiny particles—gemmules, he called them—that worked their way into sex cells. Then the gemmules originating in mother and father would blend together in offspring. Darwin, however, was wrong on two important counts. First, individual cells in the body don't cast off anything to sex cells, and, second, the units of inheritance (what Darwin called gemmules but we call genes) don't generally blend together and lose identity, but rather retain their integrity. Darwin, like virtually every other scientist of his time aside from Mendel, didn't grasp the basics of genetics as we understand it. It is a testament to the insight and sheer creativity of his theory of natural selection that it was developed in a genetic vacuum.

Once Mendel's discovery of the basic rules of inheritance was made and the term *gene* became common scientific parlance, genes quickly became thought of as *the* means by which traits could be transmitted across generations. We see this trend continuing today in research labs throughout the world as well as in the media in reports of genes for schizophrenia, genes for homosexuality, genes for alcoholism, and so on. Genes for this, genes for that. We live in a time when molecular biology seems to be front-page headline news almost every week.

If evolutionary biology as a discipline could be summarized in a single sentence, it would read something like this: Genes are selected to do whatever it takes to get copies of themselves into the next generation; everything else is just details. As such, natural selection works most efficiently and powerfully on behavioral traits directly linked to reproduction. Hence, one might think that if there is one time when genes would affect behavior most strongly, it would be during mating—and so this is the behavior that we shall focus on primarily in this book.

There is no lack of genetic models of mate choice or data that support or fail to support these models, but one thing will become obvious once we work our way through genetic models of mate choice and empirical tests of these models: Despite being the framework for conceptualizing and analyzing mating for the past sixty years and despite the fact that genes clearly play a role in how animals choose their mates (and many of their other behaviors as well), the literature on genetic control of mating obviously cries out for more sophisticated approaches that take cultural variables due to imitation into account. Genetic models of mating assume that mate choice is completely beholden to genetic predispositions. They take as a given that females and males pay no attention to the actions of others in their population and don't change their behavior according to some simple cultural dictates. We now know that this is simply incorrect. Cultural rules matter—and they matter a lot.

CULTURAL TRANSMISSION OF INFORMATION

Genes make copies of themselves and get those copies into the next generation. They affect behavior and, more particularly, mate choice behavior. They are a robust transmitter of information across generations. Another important means by which behavioral information can be transferred both within and between generations—a means embraced by many psychologists since the inception of their field—is the cultural transmission of information.

The notion that culture can play a role in nonhuman behavior goes back as least as far as Darwin's friend George Romanes. Romanes was one of the first psychologists to do detailed studies of animal intelligence, but for our purposes, his importance lies in the fact that he did pioneering work in the area of imitation, and his work showed that animals use techniques like imitation to transmit information.

There are literally hundreds of definitions of *culture.* The one you use depends on whether you are an anthropologist, psychologist, sociologist, or biologist, not to mention what subdiscipline you are in. Here, I view culture, and more specifically, cultural transmission of information, as involving some mix of trial-and-error learning, social learning via observation and imitation, and in special cases, teaching. Trial-and-error learning on its own does not constitute culture, because it does not include the transfer of information across individuals, which is a prerequisite for culture. For example, Ivan Pavlov, who introduced conditioned behavior to the world of psychology in the late 1800s (and won the Nobel prize in 1904), trained his dogs to learn that the sound of a bell meant food, but no information passed from dog to dog in these experiments. If one of Pavlov's dogs had taught another dog to pair the bell and food stimuli together, or even if one dog learned this from watching another dog being trained, then we would start talking about dog culture.

Culture itself can evolve in at least two different ways. First, there might actually be genes that code for cultural rules (let's call this type I cultural evolution). For example, imagine a gene with two variants (technically known as alleles). Variant 1 of our gene instructs individuals to imitate some behavior—let's say food-gathering techniques—that it sees others undertake; variant 2 does not code for imitation. If imitating others works better than not imitating, then we would expect variant 1 of the gene to start increasing in frequency through time. So the tendency to imitate spreads through a population. This can easily lead to very rapid changes in, say, food-gathering techniques—changes that take place much faster than possible if none of our genes had coded for the possibility of imitation. More important, we shall see that even if genes do code for cultural rules, changes in behavior through time can be almost completely divorced from the genes that coded for culture to begin with.

The second way that cultural transmission can evolve is quite different from that already described in that it lacks any genetic underpinning to speak of. This sort of cultural evolution, which we'll call type II cultural evolution, works in a way analogous to genetic evolution, except using cultural rules, rather than genes, as the unit passed down across generations. Cultural norms that outcompete other norms (by increasing the reproductive success of those who adopt them) can spread through time, just as genes do. For example, suppose you are in a foreign land and are faced with a problem. Imagine you can adopt one of two cultural norms: "do as you always have done when facing similar problems" or "look around and adopt the behaviors used by locals." If the latter rule works better (that is, provides more benefits), and other individuals then learn this rule by either imitation or teaching, then cultural evolution will favor our "when in Rome" rule, and it will increase in frequency through time.

There are two absolutely critical differences between genetic evolution and cultural evolution. First, unlike genetic evolution, in the case of type II cultural evolution, behaviors can spread even if they don't necessarily provide benefits to the individuals adopting such norms. This can be a complicated process. Second, and more important, there are huge differences in the rate at which genetic and either type of cultural evolution operate. When genetic evolution operates quickly, we are usually talking about hundreds, if not thousands, of generations for natural selection to make a noticeable difference in most behaviors. In other scenarios, we might be talking about tens of thousands of generations for genetic evolution to create real change. Not so for cultural evolution, which can easily have a huge impact in just a handful of generations. In fact, cultural evolution can have a dramatic impact within a single lifetime.

In June 1836, Nathan Rothschild, arguably the richest man in the world at the time, left Frankfurt to attend the wedding of his son Lionel. He developed a boil and went to various doctors, but continued with his daily business, though the boil got worse and worse. By the end of July, Rothschild was dead. It is not clear whether the boil killed him or contamination from the knives of the surgeons who tried to lance it did him in, but that is beside the point. Rothschild lived 150 years ago—roughly seven or eight human generations, or not nearly enough time for genetic evolution to affect our resistance to boils in any significant manner. But cultural evolution, with its mechanisms of education and imitation, in just the past generation has provided us with the medical knowledge to make the ailment that killed Rothschild a trivial medical problem in most places throughout the world. When it comes to the speed at which they operate, cultural evolution leaves genetic evolution in the dust.

Despite an obsession with the notion of culture and the importance of culture in shaping behavior, anthropologists and psychologists rarely develop models that make precise predictions about how cultural transmission will affect the distribution of any particular behavior over long periods of time. Such modeling has generally fallen on the shoulders of evolutionary biologists. The models recently developed by evolutionary biologists stand genetic models on their head and make all behavioral scientists rethink not only how culture operates with regard to a particular category of behavior such as mate choice, but to what extent culture can have a dramatic impact on simple animals with relatively small brains. Culture opens the door in these animals for the actions of a few to have serious consequences for the evolutionary trajectory of the many.

A number of different theoretical approaches have been employed to model the culture of mate choice or, more specifically, to model how females copy the choices of other females in their neighborhood. Certain corners of the animal kingdom capture aspects of all these theories, as we shall now see.

WHITE-BEARDED MANAKINS AND BLACK GROUSE

While studying the mating behavior of the white-bearded manakin bird in Trinidad in 1974, Alan Lill observed that a single male manakin on one breeding arena was obtaining more than 80 percent of all matings. Lill suggested that one explanation might be that females copy or imitate each other's choice of mates; that is, female mate preferences may be transmitted through a rudimentary form of culture.

Following Lill, many others have found that in birds and mammals that mate in arenas (also called leks), a single male often gets most of the matings, despite the presence of many other eligible males. Demonstrating that females were copying each other in such mating arenas, however, was very difficult, and the phenomenon remained largely anecdotal until the early 1990s, when a group of Scandinavian researchers, lead by Jacob Höglund, began a detailed study of mate copying in black grouse birds.

Surrounded by Scotch pine, Norway spruce, and birch, black grouse mating arenas are interspersed throughout the bogs of central Finland. As with manakins, a single "top male" grouse obtains about 80 percent of matings at an arena. Before mating, females visit arenas many times, and often in groups of females who stay together and synchronize their trips to various male territories at an arena. Höglund and his colleagues observed that a male that had recently mated was likely to mate again sooner than by chance, suggesting a possible role for imitation. In addition, older females mated, on average, three days earlier than younger females, suggesting that copying, if it occurred, was most common among younger females. This is precisely what one would expect, as young and inexperienced females stand to gain much more from watching old pros than vice versa.

Höglund and his colleagues undertook an ingenious experiment using stuffed dummy females placed on male territories within a lek. In this experiment, seven males on a particular lek had stuffed black grouse dummies placed on their respective territories early in the morning, before females arrived. The males courted these dummy females and even mounted them and attempted numerous copulations. Their results indicate that females were more interested in a male with other (dummy) females on his territory. This finding is precisely what one would expect if copying, rather than some set of physical traits alone, explained the well-known skew in male reproductive success on black grouse leks.

Höglund's work, along with numerous other studies, will clearly demonstrate that there need not be a strong link between intelligence and culture. Creatures a quarter of an inch long may lack intelligence, but they don't lack a rudimentary form of culture. What matters is not brain size so much as the ability to incorporate what others are doing into one's behavioral repertoire.

THE GIRAFFE'S NECK

Recent work on cultural evolution breathes new life into one of the most controversial and demonized names in the history of biology: Jean-Baptiste de Lamarck (1744–1829). Despite his substantial reputation as a major botanist of his time, Lamarck's name is linked with one wrong-headed idea: the inheritance of acquired (physical) characteristics. This concept was part of Lamarck's grander "theory of transformation" and worked as follows: Individuals can cause a change in an organ through constant use, and that change (the newly acquired characteristic) can be passed on to the next generation.

The classic example used to illustrate Lamarck's idea is neck length evolution in giraffes. Standard natural selection models posit that natural variation in neck length would result in individuals with longer necks obtaining more food. If neck length was a heritable trait, those with longer neck lengths would produce more offspring, and through time the average neck length in a population would increase. Lamarck's theory, however, works quite differently. Under the inheritance of acquired characteristics model, giraffes, through constant use, might be able to stretch their necks a bit longer with practice. This new longer neck trait—the result of constant attempts to reach food high in the trees—is then passed on to offspring. In modern terms, what Lamarck was arguing is that simply using an organ causes it to change; this change affects the genetic underpinnings of the organ and is passed on to the next generation.

We now know that Lamarck's arrow of causality was pointing in the wrong direction. Although it is certainly possible, indeed likely, that use can change the structure of an organ in an individual, it does not feed back and actually change an individual's genetic makeup. Lamarck was simply wrong about acquired traits' changing the genetic structure in a way that can be passed down through generations. But although Lamarck was incorrect about acquired characteristics and genes, he was not wrong about acquired characteristics and evolution. Cultural evolution is fundamentally concerned with the acquisition of novel traits and passing on such acquired traits to the next generation. Given that no one of his time knew anything about genes, Lamarck's intuition is far less misguided than many have been led to believe in the twentieth century.

It is worth noting in passing that despite the fact that many of Darwin's ideas on natural selection are quite different from those of Lamarck with respect to how evolution proceeds, Darwin himself accepted Lamarck's ideas on the inheritance of acquired characteristics. In fact, Darwin, although he believed in the primacy of natural selection, felt strongly enough about Lamarck's ideas on acquired characteristics that he went to considerable trouble to show that they were not contrary to his own.

THE NATURE/NURTURE ASSUMPTION

The relationship between genetic and cultural evolution is not a rehashing of the highly politicized and murky "nature versus nurture" debate. To begin with, the nature versus nurture question has always been fuzzy in the sense that the fundamental terms in this debate—*nature* and *nurture*—cannot at this stage be usefully defined. *Nature* can mean either that (1) a trait is controlled at some level by a gene or a set of genes or (2) not only is a trait under genetic control, but many different variants of the genetic trait exist. If everyone had a gene that controlled the expression of behavior X, then we would be talking of "nature" in the former sense, but not the latter (since everyone has it, there is no genetic variation).

The same sort of ambiguity that we see for the term *nature* also holds true for *nurture*. In *The Nurture Assumption,* Judith Harris shows how nurture historically was thought to be synonymous with "parental environment," while her own work argues that the environment of the peer group is really where the nurture effects kick in. Fortunately, in discussing the interaction of genes and culture today, the scientific literature on both genetic and cultural evolution allows us to be quite precise in our definitions.

A further difference between the perspective taken here and that of the nature versus nurture debate centers on the distinction between individuals and populations. Nature versus nurture focuses on the individual—for example, did someone fail (or succeed) because "nature" or "nurture" was the dominant factor? The genes and culture view presented here focuses on how individual behavior can have population-level changes over long periods of time. What's really interesting about this approach is recognizing that large-scale shifts occur because of genetic and cultural evolution in the landscape of life.

Nature and nurture are often depicted as discrete, mutually exclusive forces examined at one particular point in time. Genetic and cultural components to behavior, however, are both quite dynamic, shifting through time and capable of running away in very unexpected directions. Genes and culture can interact positively, or they may oppose one another. Either might overwhelm the other depending on conditions. None of these attributes has been part of the nature versus nurture controversy.

This is a golden age of theoretical biology, and Robert Boyd and Peter Richerson are among its leaders. In *Culture and the Evolutionary Process,* they note that biologists and anthropologists often ask, "Why not simply treat culture as a . . . response to environmental variation in which the 'environment' is the behavior of conspecifics?" That is, why not think of culture as just another means by which organisms adapt to the environment and leave it at that? Why all the hoopla? The reason is simple: cultural influences, unlike other environmental influences, are passed on from individual to individual. This means that the behavior of a single individual can potentially shift the behavior patterns seen in an entire population, all in the time span of less than a lifetime. That makes all the difference.

THE REAL CULTURE WAR

Historically, evolutionary biologists have been quite leery of adopting the notion that culture is important in understanding behavior in animals. In fact, many prominent evolutionary biologists are still leery of divorcing any behavior completely from genes, even in humans. Richard Dawkins became a giant of the history of biological science in the latter half of the twentieth century for his idea of the selfish gene, which says, in effect, that we are all gene robots. All animal behavior can be explained through the genetic imperative to replicate genes. He has argued for decades that virtually all of what we see, in both animals and humans, can be tied back to the genes. Genes, the argument goes, have deceptively long reaches, so that the abodes that animals (including humans) live in, their more complex behavioral traditions, and even their personalities, while appearing somehow to be divorced from genes, really are not. The power of the gene, so it is argued, works indirectly to produce all of these phenomena. Culture, then, is just genes working behind the scenes—and that's it. To be fair to Dawkins, he has softened his view on this point somewhat by exploring his concept of a discrete cultural unit he dubbed the *meme* (and which we shall examine in much greater detail in Chapter 5), but many behavioral biologists dogmatically hold his pre-meme views; and, after all, the meme still plays second fiddle to the gene in Dawkins's orchestra of life.

This skeptical response to the contention that culture is a powerful evolutionary force, in everything from simple to complex life forms, is in its own right due to many factors, not the least of which is the lack of a sound theoretical framework for understanding how culture can operate across long periods of evolutionary time. In the last twenty years or so, thanks to a handful of evolutionary biologists, such a framework for understanding the evolution of culture has emerged. This theoretical revolution has led empiricists studying the evolution of behavior to change their view of how behavior evolves and to design new studies to address behavior and transmission.

In addition to the emerging acceptance of two means for transmitting information—genes and culture—there have been new theoretical advances that examine when one might expect each of these transmission modes to be most prevalent. These models generally predict that genetic transmission will be most efficient in stable environments. When the capacity for culture exists, however, it will be a particularly efficient means of transmitting information about behavior in environments that are in constant change. The logic of this finding is simple. When things stay relatively stable, a fixed means of transferring information (one that does not rely on the vagaries of learning from others) will be selected; hence the success of genes in such environments. When the world around us is constantly changing, in both the short and the long haul, some means that can allow for new rules and innovations (even with the associated costs of making errors) works best. Hence, cultural transmission should be most prevalent in these environments.

INTERACTION, INTERACTION, INTERACTION

Cultural and genetic evolution interact in bizarre ways. These two forces may act in the same direction, or they may conflict with one another. Interactions in which genes or culture or some combination of the two win the day are all possible. Genes may even code for culture, but the manifestations of such culture—for example, fads and crazes—cannot be measured in any meaningful way by studying genetic architecture.

In their dual inheritance models, theoreticians Boyd and Richerson argue that all of the forces that lead to changes in gene frequencies have analogues within the realm of cultural evolution. Their models demonstrate how cultural change can be studied with techniques similar to those developed by population geneticists. Furthermore, they demonstrate how cultural evolution and genetic evolution can operate in the same or opposite directions, and how either can be the predominant force, depending on the particular scenario.

According to Boyd and Richerson's theory, examining cultural and genetic influences on mate choice independently only sets the stage for the $64,000 question: How do genes and culture *interact,* either cooperatively or competitively, to shape behavior? And *under what conditions do we expect cultural influences to outweigh genetic factors, and vice versa?* To address these issues, we will examine two very different modes by which genes and culture potentially interact. On the one hand, genetic evolution and cultural evolution may be two distinct processes that either reinforce one another or are in opposition. A more indirect and complicated means by which genes and culture may interact is that genes may actually code for cultural norms, allowing for some adaptive behaviors to spread quickly through a population. While this causes the distinction between these processes to blur somewhat, it also produces the bizarre possibility that genes code for culture and yet culture may produce behavior different from that predicted in the presence of genes alone.

Data now emerging show the fascinating and unexpected ways that genes and culture actually interact in animal mating situations. Consider the case of a fish less than an inch long: the guppy. In this species, females have an innate preference for males with lots of orange body color. Combining the importance of female mate copying with the documented genetically based preferences that female guppies exhibit for colorful males creates an ideal system in which to examine the relative importance of genetic and cultural factors in shaping female mate choice. In a 1996 experiment in my lab, I did just that. Essentially I created an evolutionary soap opera. A female's genetic predisposition was "pulling" her toward a more orange male, but social cues and the potential to copy the choice of others was tugging her in the exact opposite direction—toward the drabber of two males. When males differed by small amounts of orange, females consistently chose the less orange males. In other words, they copied the choice of a female placed near such a male. Here, culture—in this case, the tendency to copy mate choice—overrode a genetic predisposition for orange males. If, however, males differed by large amounts of orange, females ignored the choice of others and preferred the oranger males—in this case, genetic predisposition masked any cultural effects. With guppies, it is as if a threshold color difference exists between males in the eyes of female guppies. Below that threshold, cultural effects are predominant in determining female mate choice, and above that threshold genetic factors cannot be overridden—and this in fish with a brain the size of a pinhead!

In order to examine how genes and culture work together, we will begin by examining genetic and cultural elements in isolation. And to see why "selfish gene" models have made a major contribution to our understanding of the evolution of behavior, but fail to capture the importance of cultural evolution in animals and humans, it is necessary to devote the next chapter to the genetic models that have been used to describe the way individuals select mates.

2

Genetic Love

How do I love thee? Let me count the ways.

ELIZABETH BARRETT BROWNING

IT IS AMAZING HOW MUCH MILEAGE ONE CAN GET from creating just the right metaphor. Metaphors certainly have their place in science, particularly when communicating technically complex or philosophically detailed arguments. The undisputed champion of creating biological metaphors is Richard Dawkins. No other metaphor in biology has taken hold and entrenched itself in the minds of biologists over the last twenty plus years more than Dawkins's "selfish gene." First introduced to the scientific community in detail in *The Selfish Gene* in 1976, this metaphor is now so common as to be tossed around in all academic disciplines, including the social sciences (although there it is often mentioned with disdain).

As Dawkins himself makes clear, genes aren't "selfish" in any emotional or morally questionable sense. In fact, they aren't anything but a series of tiny bits of DNA strung together in a particular sequence and orientation, and somehow delineated from other such sequences of tiny bits of DNA. But genes appear to be selfish because natural selection favors those genes that are best at securing the most copies of themselves in the next generation. Sometimes this entails ruthless competition that produces genes that are coded to do whatever it takes to obtain the holy grail of representation in the next generation. So natural selection often, but not always, produces genes that appear to be selfish in the sense that they are programmed to do only that which helps themselves. There is no arguing the utility of the selfish gene metaphor. I use it all the time because it captures an important component of evolutionary biology.

The trouble with the selfish gene concept is that many people believe that the principles underlying this metaphor explain *everything* about animal behavior, and a lot about humans to boot. I will be challenging that assumption by presenting the case that cultural transmission and gene/culture interactions are serious, underestimated forces in evolutionary biology. This fact has been most clearly demonstrated in the context of how animals choose their mates. Yet virtually all the theoretical and empirical work on mate choice in animals uses selfish gene–like ideas as their starting point.

Selfish gene thinking is entrenched in general, and in

mate choice work in particular. As we shall see, the selfish gene perspective misses much of the big picture. Only by understanding the dynamic aspects of cultural and gene-culture interaction in mate choice can we truly understand how we and other animals choose our mates—and, by extension, how decision making evolves in general.

A BRIEF HISTORY OF COURTSHIP

Speculation on the nature of sexual attraction abounds. We could consider cafeteria conversation, television drama, folk music, or even shag pile carpet, but let us return to Charles Darwin. In his *The Descent of Man and Selection in Relation to Sex,* Darwin puts forth his now-famous theory of sexual selection to explain why individuals end up with the mates they do. Darwin proposed that animal life in one way or another boils down to one big competition for mates. In most species, males compete most directly for females, not vice versa. Basically, this difference is due to the fact that males often produce millions of sperm, creating the possibility that some males will have extraordinary success, while females' eggs are few and far between. This creates a scenario where the eggs produced by females are a scarce commodity that is well worth competing for if you are a male.

Darwin's novel ideas about the struggle among males for mating opportunities formed the basic foundation of our current understanding of sexual selection. According to

Darwin, any male trait, ranging from horns and elaborate feathers to courtship displays, that confers mating and fertilization advantage will evolve in a population. Males with such traits will simply produce more offspring than their competitors. Darwin called this evolutionary process *sexual selection*.

Competition for females often invokes images of males fighting to the death. Such "battles to the death" do occur in the animal kingdom but are rare, with less dangerous direct competition between males being the norm. In the parlance of sexual selection, this is called *intrasexual* selection, since the competition for access to one sex (usually females) is directly between members of the opposite sex. Another common mechanism of sexual selection is female choice, wherein a female actively chooses which male, among many, she will mate with. This is often labeled *intersexual* selection, because it requires individuals of both sexes in the decision-making process. Females do the choosing, and males indirectly compete with each other to be the one chosen. Darwin proposed that many conspicuous sexual traits in males, such as ornamental plumage, bright colors, and courtship displays, evolved through female mate choice.

The idea that females discriminate and actively choose their mates was controversial from its inception. Part of this controversy is probably related to the fact that while male-male battles are often spectacular and obvious, female choice is generally a much more subtle process. Over the past twenty-five years, however, a considerable body of scientific evidence has accumulated, and the traditional focus has shifted. Our understanding of how mates are selected is undergoing something of a revolution.

HOW FEMALES CHOOSE THEIR MATES: A GENE'S EYE VIEW

Studying how females choose their mates has become extremely popular in behavioral ecology. Each new issue of the many behavioral ecology journals is likely to have at least a couple of female mate-choice studies. These studies vary dramatically in terms of the species studied, the ecological underpinnings of the system being investigated, the specific male trait that attracts females, and a suite of other variables. One overarching similarity, however, linking the thousands of studies of female mate choice is that virtually all of them either explicitly or implicitly assume that a female's choice is under some sort of genetic control. That is, they assume that a gene or a suite of genes codes for what females find attractive in males, and that variations of such a gene complex exist in any population being studied. How a gene complex codes for mate choice is rarely known, and where such genes reside is almost never known, but that such genes exist is almost universally assumed.

I have no objection to supposing that genes are involved in mate choice in many species. However, there is no reason to assume that is always the case, and when genes are involved in mate choice, they may be only a very small part of the picture. But let us begin by examining genetic models of mate choice and empirical work that supports or fails to support such models.

Genetic models of female mate choice can be broken down into four groups: direct benefit, good gene, runaway selection, and sensory bias models. I will spare you the nasty mathematics that are sometimes involved, but we shall examine each of these in turn. The question I have when reading through the monthly reports of this research in scientific journals is: How would things change in system X if cultural transmission were playing a role? But the experiments and models on genes and mate choice are persuasive in themselves and must be acknowledged.

Parasites and Chemical Gifts (Direct Benefit)

Evolutionary models don't get simpler than the direct benefit models of female mate choice. These models are essentially the application of basic natural selection thinking to the question of how females choose their mates. In direct benefit models, natural selection favors females that have a genetic predisposition to choose mates that provide them with any tangible resource (above and beyond sperm) that increases their child-rearing potential or their survival, or both. For example, females may be favored to mate with males that provide them with a safe haven or food, or a male that has somehow conveyed information that he will be a good father (that is, will provide resources for the female and her offspring).

Oddly enough, despite the fact that every behavioral ecologist accepts that females do choose males, at least in part, based on their ability to provide direct benefits, the direct benefit models have rarely been tested in a rigorous way. One possible reason for this lack of experimental work is the nature of science today. Researchers simply view direct benefits as an obvious force acting on females, thus removing the impetus for controlled experiments. This produces the paradox that direct benefit models may be the most predictive but least tested of all the evolutionary models of female mate choice developed to date. Yet it isn't as if there are no tests of the direct benefit models. Two fascinating cases focus on nuptial gifts in flies and the dangers of parasite transmission in barn swallow birds.

Walk around in the woods of southeastern Michigan in July, and you will undoubtedly notice scorpion flies (*Hylobittacus aplicalis*) all over the place. Adult populations number in the thousands per hectare. Before you start swatting these flies, consider this: these relatively primitive insects are actually engaged in the hauntingly human activity of presenting one another with courtship gifts. It is these courtship gifts that have been examined by Randy Thornhill, a leader in the work on insect mating systems, to test the direct benefit models of female mate choice.

Female scorpion flies have a simple rule: do not mate with a male unless he brings you a nice, big, juicy prey item to feed on. From the male perspective, this poses a problem: hunting such tidbits proves to be quite dangerous, and at any given time only about 10 percent of males are in possession of prey that might qualify as nuptial gifts. Courtship involves a male's presenting a nuptial gift to the female and mating with her as she eats it. Males that bring no prey are immediately rejected. But female discrimination abilities are more finely tuned than that: they actually mate longer with males that bring larger gifts. If a male brings a gift smaller than a critical mass (16 mm^2), females will mate with that male for about 5 minutes, as opposed to the 23-minute matings that typically occur when prey are larger than the critical mass. Thornhill discovered that matings that last 5 minutes or less often involve no sperm transfer. That is, if males don't bring a big enough prey item, they may get a chance to mate for a few minutes, but they shouldn't expect many offspring to come of it.

Given such discrimination abilities on the part of females, it seems more than reasonable to ask what *they* get out of the whole process. Females can clearly distinguish males with large gifts from cheapskates, but what benefit do they receive by doing so? It turns out to be something of a bonanza. Females that actively choose males that bring large nuptial gifts produce an increased number of eggs and, in all likelihood, have a longer life span. In the currency of evolutionary biology, one can hardly imagine a more direct benefit.

Male scorpion flies can't fake the quality of the mating

gift, since it is the prey item, and the prey item alone, that they give to females. In other species, however, males wrap up their prey item in a kind of cocoon before they present it to females, and in even other systems, males present the cocoon wrapping with nothing inside. In the last case, males try the equivalent of passing off a bit of cubic zirconium as a diamond by spinning a large cocoon, placing nothing in it, and attempting to fob it off on a female.

Our second example of direct benefits brings us to the farmhouses of Denmark. The Danish have a 2,000-year relationship with the fascinating barn swallow (*Hirundo rustica*), which not surprisingly often lives high up in the roofs of barns and other lofty abodes. For the last fifteen of those 2,000 years, nobody has studied barn swallows with a greater vigor than Anders Pape Møller. Møller's fascination with barn swallows is in part due to the fact that they "are almost unique among common European bird species by being sexually highly dimorphic in one character: tail length." In other words, males and females have very different-sized tails, with male tails being much longer, but in every other way the two sexes look similar.

Unless you are an ornithologist with a penchant for barn swallows, it may be hard to believe how much work has been done on barn swallow tail length, sexual selection, and female mate choice. Male tails have been cut off and other males have had extra feathers artificially sewed on, all in the name of science. And although there are dozens of studies of barn swallows and sexual selection (most by Møller), we shall focus here on barn swallows, parasites, and the selfish gene–based direct benefit model of female choice.

Although a barn is a rather cushy circumstance as far as natural populations of birds are concerned, life as a barn swallow is not all that pleasant. One unpleasant reality of life in the wild is that animals are in a never-ending struggle with parasites that want nothing more than a free, continuous supply of food. Barn swallows have their fair share of parasites, and they come in two forms: ectoparasites, which hang on to the outside of a host and easily move from one host to another, and endoparasites, which reside inside a host and require more elaborate mechanisms for moving to new hosts. Ectoparasites provide an interesting test of the direct benefit model.

In his work on barn swallows, Møller observed that females prefer to mate with males possessing elongated tail feathers. Such males also happen to be less infected with ectoparasitic mites than their short-tailed compatriots. To determine if females preferred longer-tailed males partly as a consequence of a direct benefit of mating with males with reduced ectoparasite load, Møller ran the following experiment. He fumigated a series of nests and introduced extra parasites into a different group of nests. The results of this experimental manipulation were dramatic: fumigated nests had much higher success (measured by the number of eggs reaching hatching stage) than nests with experimentally increased mite load. In addition, the offspring raised in fumigated nests were much healthier (as measured by body weight). So females with genes that favor mating with males bearing a low parasite load would receive a significant direct benefit from doing so: more and healthier chicks. Creating a parasite-free home environment (by mating with the right male) clearly has its merits.

Brave Guppies and Smelly T-shirts (Good Genes)

Direct benefits are not the only thing females get from males. In addition to obtaining such direct benefits as food, safety, and the like, females also get something quite valuable from males: sperm. One good sperm may be worth a whole heap of direct benefits. Females that mate with males possessing a suite of favorable traits—traits that can be inherited by offspring—may hit the jackpot. If such traits as size, fighting ability, and physical attractiveness are passed down across generations, choosing males with "good genes" becomes not just a luxury but a necessity.

Good-genes models apply to mating systems in which females receive *sperm and sperm alone* from males. There are many forms of the good-genes model, and they all face two hurdles. First, how do females know which males have "good" genes and which don't? It is easy to tell which male brings back more food to your nest, but how on earth do you assess a male's overall genetic quality? Second, why don't males cheat by faking the traits associated with good genes? If females choose a trait to determine which males are those with the best genes, wouldn't natural selection favor males investing in such traits, no matter what their genetic quality? Amotz Zahavi's "handicap principle" addresses both of these hurdles to the good-genes models.

The handicap principle focuses on the notion of honest advertising. Imagine that you are contemplating purchasing an item, let's say a car, and there are lots of brands from which to choose. The honest advertising principle suggests that you should distinguish among brands using criteria that really reflect something important about cars; what's more, you should use only that subset of criteria that are very difficult to fake. So, for example, you might give serious weight to how well a car's airbags work during a collision rather than what color the airbags are—the former being both important for safety and hard to fake. With respect to mate choice, Zahavi's handicap principle suggests that only traits that are true and honest indicators of male genetic quality should be used by females choosing mates.

One general characteristic of an honest indicator trait is that it is "costly" to produce and maintain (in fact, the cost of such traits is why they are referred to as handicaps). As a consequence, such a trait is then hard to fake (since it requires a significant investment to produce) and thus is expected to be associated with vigorous males. Let's once again consider parasites, except now let's focus on endoparasites—the type that live inside a body and are not able to hop from individual to individual (as in the barn swallow case). Females who choose males with very few parasites probably may not gain as much in the way of direct benefits as in the barn swallow case, but they may receive indirect benefits in that they are mating with males with "good genes"—in this case, genes that confer parasite-resistance abilities. From a female's perspective, such parasite-resistance genes could only be good for her offspring to inherit.

The catch is still, of course, how females gauge whether a male possesses good genes. For parasite resistance, the argument goes as follows: Only high-quality males can resist parasites (inferior males cannot fake this trait), and so females should choose mates based on cues associated with parasite resistance. One such cue appears to be body coloration: infected males tend to have much drabber colors than healthy males, who often are very colorful and eye catching. Studies involving dozens of species of both birds and fishes have often yielded results consistent with the handicap hypothesis in that females often choose the most colorful (and least parasitized) males.

The handicap principle is far from limited to the case of parasite resistance. Consider an experiment that a colleague of mine, Jean-Guy Godin, and I did a few years back. Godin and I had long worked together on the evolution of antipredator behavior in guppies and realized the system we had developed was ideal not only for examining antipredator behavior, but also for testing Zahavi's handicap principle.

When a fish predator is located near a school of guppies, males (as well as females) often cautiously approach this potential threat and "inspect" it by moving toward the predator and obtaining various types of information. Behavioral ecologists have suggested that bold males who dare to approach a predator may, in fact, be *advertising their overall quality* to nearby females. We examined whether this was the case for guppies; if it was, "inspection" behavior would likely be a reliable indicator of fitness and a good candidate for the handicap version of the good-genes model.

To test our ideas, we had an ex-engineer from NASA build us a custom-made experimental apparatus that allowed us all the control we needed to test our basic ideas. Our first observation, based on some work from a different experiment we were conducting, was quite straightforward: more colorful males tended to be bolder and inspected the predator. What we found next, however, was anything but straightforward. Colorful males inspected more *only* when females were in the vicinity. That is, colorful males were more likely than drab males to inspect predators, but only if an audience of females was present. Take away the females, and the differences in male inspection behavior disappears. This certainly suggested to us that we were on the right track. If inspection was costly, according to the handicap principle, those costs should be paid only if females were watching. There is no sense paying the cost if the females one is trying to impress are absent.

The big question that Godin and I then needed to address was whether females in fact used male boldness (measured by inspection behavior) as a cue for male vigor and subsequently chose such bold males as mates. What we found was that females that were allowed to watch males inspect a predator were much more likely to mate with bold males than with timid males. So females were judging males as potential mates on something. We then needed to ascertain exactly what that something was, for remember that bold males also tend to be colorful males, and we still needed to determine whether boldness or color was the handicap trait that females used in selecting their mates. This is where our custom-designed experimental tank paid off.

The tank we had designed for these experiments allowed us to simulate inspection and thus separate inspection and color as the potential driving force in this system. Through the use of tiny Plexiglas tubes that contained a minuscule mirror that kept guppies oriented toward the predator, we could take a male and control his inspection behavior. Using a pulley system, we could make colorful or drab males inspect a predator closely or not at all. The results were conclusive: females were interested in bold males as mates regardless of the male's color. Boldness was the costly signal that couldn't be faked and presumably reflected a male's overall genetic quality. (Male readers may wish to think about that the next time they decide to go skydiving.)

Not all good-genes models rely on males' possessing a costly handicap. Recall that all that is required for such models to operate is that males differ in the quality of genes they possess and that females are able to distinguish among males based on such differences. Handicaps make that requirement easier to meet, but they are not prerequisites. A recent experiment on mate choice in humans (which was dubbed the "smelly T-shirt experiment" by the *New York Times*) illustrates how good-genes models can operate without handicaps.

Evolutionary biologists have argued for some time now that a suite of genes called the major histocompatibility complex (MHC) may play a role in both animal and human mate choice. The MHC suite of genes is involved in disease resistance; the proteins coded by such genes help the body identify if something is "self" or "foreign." What makes the MHC so fascinating is that it is one of the most variable suites of genes ever uncovered. Very few (if any) individuals have exactly the same MHC. One hypothesis is that individuals should prefer mating with others who have a dissimilar MHC, because offspring from such a mating will have a new combination of MHC genes. Such new MHC combinations might be particularly good at providing immune system protection, the basic idea here being that diseases evolve very quickly because they reproduce at such high rates. One way to fight newly emerging variants of a disease is to have a constantly changing immune system—hence, the preference to mate with individuals with a dissimilar MHC.

If it is true that choosing a partner with a dissimilar MHC is akin to choosing someone with good genes, how is it possible to determine who's who? Work on rodents provides a hint. Although we don't know exactly why, it turns out that mice and rats can use odors (in the rodent case, the odor of urine) to determine if another individual is a good MHC match. Claus Wedekind, of Bern University, reasoned that humans too might benefit by an appropriate MHC match and that we too might use odor as a means for searching out the best mates.

Wedekind gathered up a crew of undergraduate males and females and undertook the "smelly T-shirt" experiment. Forty-four men were told to wear a cotton T-shirt for two nights and to make sure to stay away from any strong odors during this time. Wedekind also took a blood sample from each male to characterize his MHC. On the other side of the coin, forty-nine females also had blood samples taken for MHC analysis. These women were then given T-shirts from males with similar MHCs and males with dissimilar MHCs. Wedekind found that women (who were not taking oral contraceptives) consistently found the odors of the T-shirts from males with dissimilar MHCs sexier.

Long Eyes and Orange Spots (Runaway Selection)

In runaway models of animal mate choice, genes in males and genes in females become "linked" to one another. To see the basics of such a model, imagine that some females in a population have a genetic predisposition to prefer a certain characteristic in males. Some males in this population have the trait that females prefer; some don't. In runaway models, a genetic correlation is posited between genes controlling two different traits: the particular trait in males (a trait that females prefer) *and* the mating preference in females. For example, females that prefer to mate with large males should produce not only large sons but daughters that possess a genetically coded preference for large males. After many generations of such matings, genes that code for a preference for large males in females and the genes for large body size in males become linked, and are no longer independent; when one changes, the other changes as well. Once this phenomenon is established, it takes off ("runs away") like a snowball rolling down a snowy mountain. A positive feedback loop is in place, and natural selection may produce increasingly exaggerated male traits and stronger female preferences for these exaggerated traits. Such exaggerated traits are often striking to the animals involved, as well as to any human observers. They are the stuff of nature documentaries, yet few outside the science of evolutionary biology have been taught the underlying theory (runaway selection) that produces such traits.

The best demonstration to date of runaway genetic selection comes from Gerald Wilkinson's work on the stalk-eyed fly *(Crytodiopsis dalmanni)*. In this rather odd species, females prefer to mate with males possessing eyes that are at the end of long eye stalks. For thirteen fly generations, Wilkinson chose males with long eye stalks and allowed them to breed with females in one treatment, while creating another group in which only males with short eye stalks could breed. Not surprisingly, in the line where long eye-stalk length was selected, the average eye stalk increased in length and a more and more exaggerated, longer version of the male trait was seen. Conversely, in the line where short eye-stalked individuals were selected, the size of the eye stalk decreased—a solid, but not unexpected, result, and the sort of thing that hundreds of experiments had shown natural selection capable of.

What is truly important about the experiment in terms of runaway gene models is that Wilkinson found a positive link between the length of male eye stalks and female preference for this male trait. Females from the long eye-stalk line preferred males with long eye stalks, and females in the short eye-stalk line preferred males with short eye stalks. This was the case despite the fact that Wilkinson was randomly choosing females throughout the experiment. There was absolutely no selection for any type of female based on what sort of male she preferred as a mate (short eye stalk, long eye stalk, or something else). The female preference for male eye-stalk length thus changed as a response to genetic changes in males, in accordance with the prediction of the runaway model.

More evidence of runaway sexual selection comes from the guppy. Female guppies from the Paria River in Trinidad have a strong preference for males with lots of orange color on their bodies. Given this strong preference and the fact that orange color in males is passed along from father to son, Anne Houde sought to discover a link between the genes that code for orange in males and the presumed gene(s) underlying female preference for mating with such males.

In a classic artificial selection experiment, Houde created four high–orange color groups (containing males with lots of orange) and four low–orange color groups. Males in each group were allowed to mate with females. In each generation, Houde selected the brightest orange male offspring in the high-orange groups and the drabbest males in the low-orange group. Selected males would be allowed to breed with randomly chosen females, and this would go on for four generations in both high- and low-orange groups. At the end of generation four, Houde not only found that males in the high-orange group had evolved to be even brighter (and the converse in the low-orange group), but that females in the high-orange group showed a stronger preference for brighter males as well.

Houde's results certainly suggest that runaway selection might be operating in the guppy system. That being said, when Felix Breden and Kelly Hornaday ran a similar experiment on a different population of guppies, they found no evidence of runaway selection, dampening the enthusiasm that the guppy system was one of the few that had strong data supporting the runaway model.

Of Swords and Chucks (Sensory Exploitation)

The basic premise of the sensory exploitation model is that genes that code for some preference in one context are shanghaied into the realm of mate choice. Michael Ryan and John Endler, the most vocal proponents of this view, suggest that females may have "preexisting biases" for certain male traits and that such biases are then exploited by duplicitous males.

The sensory exploitation argument goes as follows: Suppose that, for whatever reason, red berries are the most nutritious food source available to a fruit-eating songbird species. Females that are best able to search out and subsequently eat red berries survive and reproduce better and are thus selected over females that eat other types of berries. If red feathers should suddenly arise in males of this normally blue-feathered species, they may be chosen as mates because the female's nervous system is already designed to respond preferentially to red objects. Males with red feathers are exploiting females in the sense that such males are favored not because the trait they possess means anything (it may have no fitness consequences); they simply attract the attention of females.

One of the most convincing cases of the sensory exploitation of female mate choice is Alexandra Basolo's work on two related species of tropical fish, the green swordtail (*Xiphophorus helleri*) and the platyfish (*Xiphophorus maculatus*). Female swordtails prefer males with long "swords" (a colorful elongation of the tail fin). Some male swordtails have long swords, and some have short swords, but platyfish males lack swords completely. To examine sensory exploitation experimentally, Basolo sewed artificial (plastic) swords on platyfish males and examined how female platyfish responded to this new twist of fate. What she found was that female platyfish showed an immediate, strong, and consistent attraction to males with newly acquired swords over naturally "swordless" males. Female platyfish must have had a preexisting bias for long tails, because despite the fact that there had been no evolutionary history of choosing males with swords (males in nature never have tails!), this elaborate male trait was viewed as very attractive as soon as it appeared in the population.

The second example of sensory exploitation comes from two species of frogs: *Physaleaemus pustulosus* and *Physaleaemus coloradorum*. Males in these species use advertisement calls to attract females, and, as is the case for most *Physaleaemus* species, males of both the *pustulosus* and *coloradorum* species begin such an advertisement with what is referred to as "whine." *Pustulosus* males, however, are unique in that most add a "chuck" sound to the end of their call. When females choose between *pustulosus* males that chuck and those that don't, they prefer the former. The "chuck" part of the call is and has always been absent in the *coloradorum* species. Yet when modern acoustic gadgets are used to add a "chuck" call on to the end of *coloradorum* male calls, such calls are immediately preferred by females. Once again, the preference for a more elaborate call seems to have predated the actual appearance of the elaborate call, suggesting that female nervous systems cue in on such calls instantly, once they appear.

THE MISSING PIECE

Although each example in this chapter is shown to support a genetic model of female mate choice, in no case do we actually know where the putative set of genes for mate choice resides. This happens to be true not only for the examples I chose, but for virtually every example that tests genetic models of mate choice. At one level of explanation, this observation is anything but shocking. After all, one can study the effect of something without knowing the physical location of it. In fact, much of the work on whether a trait has a genetic underpinning has been done in the complete absence of knowledge about whether the gene that causes something is, for example, located on chromosome 7 or somewhere else. If a gene has certain effects, one can study how those effects are passed down in a population without using molecular genetics to tell us exactly what the gene looks like or where it is located.

The problem with genetic models of mate choice is not that we don't know where the genes can be found. My discomfort with such models is not even that they fail to capture important aspects of mate choice. In many instances, they do. Realizing what is wrong with understanding behavior through genetic models begins perhaps with recognizing that while genetic models are usually created without any thought of how cultural evolution might have shaped the behavior in question, cultural evolution can potentially produce virtually the same pattern of results. Consider the following example.

For much of the 1980s, runaway models of female mate choice were the rage among mathematical evolutionary biologists. After a while, the theory was so far ahead of the experimental work that it was viewed as extremely important when the stalk-eyed fly and the guppy studies we spoke of earlier appeared and seemed to support runaway models of female mate choice. Much time and effort had gone into the development of these models, and people were excited that the predictions of the models were holding up under experimental scrutiny. The catch is that right around the same time as the guppy and stalk-eyed fly data were hitting the presses in 1994, some new models of female choice, based on the notion of cultural transmission of behavior, were also appearing in journals. These models made some of the same predictions that runaway genetic models made, but without the direct genetic underpinnings that were the core of classic runaway models. Thus, cultural transmission of mating behavior can also lead to what looks like runaway selection. Just when some data supporting the predictions of a gene-based model appeared, biologists had to admit that cultural transmission of behavior could have produced the exact same sort of behavior.

If the results of classic runaway models of female mate choice are called into question by models of mate choice that consider culture, what about the other genetic models of mate choice (direct benefits, good-genes, and sensory bias models)? Couldn't some sort of cultural transmission of behavior mimic these processes as well? Cultural transmission could be on the verge of toppling many genetic explanations of behavior.

3

Guppy Love

Birds do it, bees do it, even fish do it?

ALTHOUGH AS A TEENAGER I HAD NO IDEA THAT science would play such a large role in my future, I nonetheless possessed all the stereotypical traits of a scientist-to-be. That is to say, I was a nerd, and I never even came close to getting a date in high school. My lack of dating experience, however, was not for want of trying. Although I sadly failed to obtain any actual firsthand experience in the high school dating world, I watched others excel.

One thing I noticed was that being successful in the world of high school dating was like being successful in business: you have to get your foot in the door. Date one person, and others somehow suddenly become interested in you. People seemed to imitate the choices others make. This may not have been the case for the handsome quarterback who needed no help, but when Joe Average got one date and then other females started talking about him in a new light, something was most definitely afoot. One version of this phenomenon is seen in the folk wisdom that says once a man puts on a wedding ring, he instantly becomes more attractive in the eyes of women. It's as if desirability is a tiny snowball that stays tiny until it gets a little push down the mountain.

Years later, when I was in graduate school and somewhat recovered from the shell shock associated with entering the dating world, I began to realize that the phenomena I have described—the impact of copying on mate choice—had been all but overlooked by those studying animal social behavior. At that point in my fledgling scientific career, I was focused on studying the evolution of cooperation and altruism and had never done any work on animal mate choice. In fact, I was set against doing such studies because "mate choice" seemed to be the popular topic in behavioral ecology in those days, and I was too proud to jump on a bandwagon. People were studying every conceivable trait influencing mating habits in everything from bacteria to gorillas. I wasn't interested in tossing in my precious two cents.

The more I thought about how blatantly obvious it was that imitation plays a large role in human mate choice, however, the more I realized that this force must play a large role throughout the animal kingdom. To my surprise, when I searched the rather extensive literature on animal mate choice in 1989, there was not a single well-controlled study that examined the relationship between imitation and mate choice in animals. Here and there, if I searched hard enough, I found an anecdote suggesting that many animals might be copying each other's choice of mates, but there was nothing concrete. At that point, I realized I was going to take on the work involved and see what would come of this hunch.

COURTING A GUPPY

Despite the fact that we had a guppy tank in our New York City apartment throughout my childhood (it was one of the few pets allowed in our apartment complex), I never thought much about guppies before starting graduate school, and whatever thoughts I had were certainly not about conducting behavioral experiments on these critters.

When I started contemplating experiments on imitation and mate copying, somehow it seemed obvious to begin by reading some material on guppies. After all, I knew that guppies were wonderful for laboratory experiments in behavioral ecology. These fish are readily available, take nicely to a life in the lab, happily mate while being observed by scientists, and breed very quickly—a winning combination for a behavioral scientist. I was pleasantly surprised to find that guppies were the "white rats" of mating experiments, and that there were already dozens of papers on the sexual habits of this species. Reasoning that a good understanding of the biology of mate choice in the species I was going to examine could only be helpful in the long run, I settled on the guppy for my work. What's more, I knew that if I found that imitation played a role in mate choice, I would want to know how large a role it played in relation to genetics. Given that much was already known about genetics and guppy mating, guppies seemed the ideal species for some future experiment designed to tease apart genetic and cultural forces shaping mate choice.

In 1990, I began the first controlled study of imitation and mate choice in animals. The experiment generally consisted of two females and two males. One female (let's call her the observer) was able to view a second female (let's call her the model) near one of the two males present. The questions I addressed were the simplest ones I could think of. Was the observer likely to choose the same male as the model, and was this a direct result of her observing the model's choice of mates?

I could take advantage of the fact that guppies get used to laboratory tanks very quickly. In my first experiments, I "staged" an initial mate choice, and then tested how that affected an observer's choice of mates. The experimental apparatus I used included a 10-gallon tank. In the center of this tank was a cylinder I could raise up and down, and this cylinder was where the observer female was located at the start of a trial. Next to the 10-gallon tank at each end was a smaller Plexiglas cube that held one male. So the observer saw a single male at each end of the tank. I admit I am a little too proud of the next bit: the staged part of the experiment.

Inside the tank containing the observer were two clear partitions; one partition was about 2 inches from the male on the left side, and the other was about 2 inches from the male on the right. These partitions allowed me to choose which male a model would choose. Before the start of a trial, I flipped a coin. If it came up heads, the focal female would be placed near the male on the right; otherwise, she would be placed near the male on the left. After viewing the model near one male for a bit, the model would be removed, and I would see which male the observer chose.

Staging the experiment as I did was absolutely critical to avoid a potential confounding factor that might have rendered any experiment I conducted on imitation and mate choice useless. Suppose that instead of placing the model near one of the males based on the flip of a coin, the model was allowed to swim freely and choose on her own. Wouldn't that be more natural? Undoubtedly, but to see the problem with the more natural version of the experiment, let's say that doing the experiment the natural way I found that observers went to the same male that the models were near. What then would be the interpretation of these findings? There are at least two possibilities:

1. The observer copied the mate choice of the model.
2. The model and the observer *independently* chose the same male.

Explanation 1 is clear, but let's look at explanation 2. Despite my efforts to match males for as many traits as possible, one male may simply have been more attractive than the other, and both females may have come to that same conclusion. No mate-choice copying, just a couple of females making the same choice independent of one another. But placing the model near one male based on the flip of a coin assured me that even if males differed in attractiveness, sometimes, by chance, the model would be near the more attractive male and other times it would be near the less attractive. Tossing the coin introduced randomization, and thus removed male attractiveness as an explanation for any of the results uncovered.

The results were very clear. In seventeen of twenty trials, the observer female chose the same male as the model female did. I was convinced that I was on to a system that might be the first to demonstrate experimentally imitation in the context of mate choice, but I also knew that the coin-tossing trials were only the first step. Although these trials ruled out that females were independently picking the same males, there were other concerns. That is, although those trials were consistent with the notion that females copied each other's choice of mates, more control experiments would need to be run to rule out alternatives and show definitively that what I had was mate copying.

In the end, five subsequent control experiments were needed to rule out alternative explanations to female mate-choice copying. Let me just mention one to give a flavor for the sort of controls to which I am referring. Consider for a moment what other phenomenon could have been responsible for producing the results I found in the coin-tossing experiment. Guppies, like many other animals, live in groups and prefer to be in the company of many other fish. Larger groups are safer and provide other benefits to their constituents as well. So maybe what I thought was mate copying in the coin-tossing experiment was really just the observer's attempt to place herself in an area that was likely to have the largest group of fish in the vicinity.

Recall that in the coin-tossing study, on one side of the tank is a lone male and on the other are a male and the model female. While it is true that the model was removed when the observer made her ultimate choice, it might very well have been that the observer went to the male that was near the model *simply because that was the side of the tank that just a moment ago had two fish (versus one).* The observer's behavior rule may just have been to go "where the largest group is likely to be found" rather than "to copy the mate choice of the model." Let's call the former the group-size hypothesis and the latter the mate-copying hypothesis. Either might explain the results of the coin tossing; in order to distinguish between them, one must run a control experiment.

There are various ways one could contrive to test the validity of the group-size hypothesis. I opted to place females, instead of males, in the small cubes juxtaposed to the tank containing the model and observer females. So now all four fish in a given trial were females, but everything else was exactly as in the coin-tossing tests. If the observer went to the side of the tank where the model plus one of our new "females in a cube" were, then the group-size hypothesis was supported, as we have more evidence that it was simply having two fish on one side and one fish on the other that explains everything in our first experiment. If the observer female chose randomly between sides, the group-size hypothesis could be rejected, and the mate-copying hypothesis would receive an indirect boost. What I found, to my delight, was that observer females chose randomly.

After running four other controls and ruling out a series of alternative hypotheses, it became clear that mate copying was the only result consistent with all the work I had done. Now that I had a system in which cultural transmission affected mate choice, the door was open to looking at why animals would rely on imitation in choosing mates (what the costs and benefits of such a decision are), and just how important cultural transmission was in the big picture.

In order to examine the costs and benefits of mate copying in depth, I relied on a technique that had worked for me: I thought about imitation and mating in terms of my everyday interactions with people. In human culture, it is not too controversial to suggest that most infants, teenagers, and even young adults look to their elders for models of behavior. Older people don't tend to look at young people in the same way. So perhaps younger female guppies copy older females more than older females copy younger fish.

It is worth noting that just suggesting the use of my everyday experience as a means for devising experiments in animals would no doubt infuriate some of my colleagues. Some might accuse me of being "specieist" (yes, there really is such a thing!) by relying on human experiences to study nonhumans, instead of trying to view the problem solely through the lens of the animal under study. Others might say that humans are so different from nonhumans that my train of thought could never lead anywhere worthwhile. I dispute both objections. Although a researcher always needs to pursue objectivity, it is probably impossible, and certainly not advisable, to divorce science from personal experience. If I had done that, I'd have been able to force myself to forget the guppy tank in my childhood apartment and my unimpressive dating record in high school, in which case I might never even have thought of the original imitation and mate-choice experiment.

The protocol for the age/imitation/mate choice experiment was straightforward. In one treatment, young but sexually mature females were the observers, with older, more experienced females serving as the models. In the other treatments, the tables were turned, and the older female observed the younger. What my colleague Jean-Guy Godin and I found was that when young, impressionable females were the observers, they copied the mate choice of others. On the other hand, when the experienced lot observed the youngsters, they went about their mate-choice activities regardless of whom they saw the younger females choose. One benefit, then, of mate copying is for young females to learn who is potentially a good mate and who isn't—not by taking the time, trouble, and chances associated with assessing males, but by copying the choice of a more experienced female. Human analogies abound.

With a basic sense of some of the costs and benefits of imitation now at hand, the next step was to gauge how important imitation was in shaping mate choice. Many studies have listed variables that affect mate choice in the guppy, including color, place of birth, tail size, parasite load, and swimming speed. What I wanted to know was whether I had simply uncovered yet another small piece of the puzzle or a powerful force that is fundamental in shaping sexual behavior. I worried that mate copying was merely characteristic 29 on the list of things that affect mate choice in guppies.

One effective way to know how to rank a list of variables is to pit them against one another and see which emerges victorious. In the context of mate choice, for example, I could pit size against color. Given that female guppies like males with few rather than many parasites, I could, for instance, place another (model) female near a parasitized male and leave the unparasitized male alone. If the parasitized male became more attractive in the eyes of an observer, then imitation could be said to be more important than parasite load. But there are a lot of factors that influence mate choice in guppies, and the last thing I wanted to do was a long series of experiments that pitted mate copying first against factor 1, then against factor 2, and so on. In order to avoid this, I needed to run an experiment that stacked the odds against me and pitted imitation against everything else at once.

In 1991, again in collaboration with Jean-Guy Godin, I ran what I call the reversal experiment. This experiment consisted of two treatments. Treatment 1 had a single female and two males. Here, we simply noted which male was chosen by a female, and then 30 minutes later, using the exact same fish, we tested the female a second time to see whom she preferred. In essence, we were examining an important aspect of behavior: consistency. We were pleased to find that females were very consistent in their choice of mates; in sixteen of the twenty trials we ran, females chose the same male on both occasions. We don't claim to know precisely what characteristics females used in choosing males in this experiment, but we do know that whatever they used the first time, they continued to use it.

Treatment 2 was identical to treatment 1, with one small but critical difference. Now after a female chose between males for the first time, we gave her the opportunity to see another female choose the male she did *not* choose. The social information a subject received from the model was *always* in direct opposition to her own choice of males. Whatever bevy of traits combined to create a preference for one of the two males was now being countered by the possibility of imitation. Which male should a female guppy choose: Would she be consistent or swayed by the opinions of others? We discovered that a significant proportion of the time, female guppies imitated the behavior of the model and, in their second round of choices, went toward the male they did not choose during the initial round. Social cues overrode personal preferences, whatever those preferences were based on.

Guppies are not the only fish that display imitation in the context of mate choice. Using slightly different experimental protocols, female mate copying has been uncovered in medaka and sailfin mollies, as well as some cases that are a bit more controversial but nevertheless potential instances of mate-choice copying. What's more, it isn't just females that copy each other's choice of mates. Somewhat surprisingly, male sailfin mollies also imitate each other's preference for mates.

WHAT DOES IT MEAN TO HAVE WHAT SHE JUST HAD?

Although I hope I have made clear what constitutes mate copying in nature, I have not yet provided a formal definition of mate copying. The first formal definition of female mate-choice copying was provided by Stephen Pruett-Jones in 1992. Pruett-Jones, who often studies fairybirds in the wilds of Australia when he is not busy writing papers, argued that mate copying can be defined as the increased probability that a male will mate in the future, given that he has recently mated. More specifically, if a male has an X percent chance of mating tomorrow if he has not recently mated, and a Y percent chance if he has recently mated, mate copying is defined as the difference between Y and X. From a scientific point of view, one nice feature of this definition is that probabilities can be measured, and hence a mate-copying hypothesis can be supported or refuted by available data. However, Pruett-Jones's definition fails to address a critical aspect of mate-choice copying: individuals need to observe others choose mates for mate-choice copying to have occurred. Without observation of the mate choice of others, mate copying hasn't occurred.

We must amend Pruett-Jones's definition by adding the following sentence to the end of it: "Further, the information about a male's mating history (or some part of it) must be obtained by the female via observation." This definition retains the measurable properties of Pruett-Jones's, but stresses the importance of actually seeing what is going on.

No definition is ever perfect. Imagine a species that in its evolutionary past met my conditions for mate-choice copying. Long ago, females observed a male territory holder mating with other females and were then more likely to mate with that male as well. Suppose that females then stayed on a particular territory after mating, and a new rule of thumb sprouts up: a rule that instructs females to look at how many other females are on a male's territory. Because this information correlates well with how many other females have mated with a male in the recent past, females can get a quick assessment of the number of others that have mated with the territory holder without actually taking the time to view such matings.

Our rule of thumb should increase in frequency since females that use it avoid costly activities, and eventually females need not observe any matings but are nonetheless in some sense mate-choice copying. Yet observation of the actual mate choice of others is not part of this scenario, and so to be conservative, it is perhaps best not to include this type of mate-choice copying in our definition.

Others will argue with this definition, as is their obligation, but it is a starting point based on some combination of biological and psychological underpinnings. With a testable definition of mate-choice copying in hand, we can now examine a few cases of mate choice that at first appear to be imitation, but after further investigation probably do not meet the definition presented above.

APPEARANCES CAN BE DECEIVING: DEER, STICKLEBACKS, AND ISOPODS

Watch the behavior of female fallow deer in Britain, and it certainly looks as if they are copying each other's choice of mates. When a female decides which male's harem she will join, she seems to be particularly fond of males with large harems. The more females in a harem, the more opportunities there are for mate copying. Tim Clutton-Brock and

Karen McComb couldn't pass up the chance to run some experiments on imitation and mate choice in this species. It's fine and dandy that guppies mate-copy, but if one could show it in a charismatic animal with fur, like a deer, then you'd be on to something. But Clutton-Brock and McComb did more than simply examine mate copying. They examined whether fallow deer (*Dama dama*) are attracted to such harems because of mate copying or perhaps because being in a large group reduces the risks associated with predation. As all readers of the children's classic *Bambi* realize, life as a deer can be tough. Only a few pages into the book, and already Bambi's mother is killed. To avoid such disasters in the real world, females join the safety of large groups.

Clutton-Brock and McComb ran two experiments. In the first, pairs of estrous females were given a choice among four arenas. Two contained solitary males, one contained a male with eight females, and one a male with more than thirty females. In the first experiment, females showed a strong preference for males with harems rather than solo males. If you were rooting for mate copying, this result was at least consistent with what mate copying would predict. The trouble, of course, is that it is equally consistent with the strength-in-numbers hypothesis.

In their second experiment, pairs of females were again given a choice among four arenas. This time the arenas contained a male with nine females, nine females with no male, a male with nineteen females, and nineteen females with no male present. Results of this experiment indicated that a female's preference was not due to copying the mate choice of others, because females showed no preference for arenas containing a male plus females over arenas containing females alone. The final nail in the mate-copying-in-deer coffin is that actually seeing a male mate does not increase the probability that an observing female will join his harem.

For a second example of what could be, but probably isn't, mate copying, let's return to fish. A number of studies of female choice in fish clearly show that females prefer to mate with males that already have eggs in their nest from prior matings. Initially, people argued that because male sticklebacks, for example, that had recently mated are also chosen as mates in the present, female sticklebacks mate copy. According to Pruett-Jones's definition, this would indeed constitute mate copying, because a male's chances of mating are clearly influenced by his mating success in the immediate past. But chances are that females making a decision to mate or not haven't observed a male mating with anyone; they have simply seen the results of such a mating. According to my definition, then, we have not met the criteria for copying mate choice.

A reasonable person, however, might say, "So what if it doesn't meet Dugatkin's definition of mate copying"? What does it matter if females didn't actually see a male mate with anyone else? The eggs in his nest are evidence that he did! Indeed, the eggs *might* be used as a sign that this male just mated, but the eggs *might* also be a signal of other things, and these other things might be what drives the female's decision process. For example, eggs in a nest may indicate that the male possesses the good genes we read about in Chapter 2. Such males may simply have good genes that allow them to be fit enough to defend a nest against predators successfully. If this was the case, any implication of mate copying would be misplaced.

Another explanation for why females choose to mate with males that have many eggs in their nest is the increased courtship rates that such males display. Males with eggs often court more, and this courtship alone may explain their increased ability to attract more females, removing any impetus to search for mate-choice copying. Yet another alternative explanation to female mate copying is the dilution hypothesis. That hypothesis states that if you place your eggs where there are already lots of other eggs, should a predator come around, your eggs stand a good chance of surviving. Suppose, for example, that predatory fish eat a thousand eggs in any nest they find. If you laid five hundred eggs in a nest that had no other eggs in it, then should a predator get to such a nest, all your eggs would be eaten. If, however, you laid your five hundred eggs in a nest that already had forty-five hundred eggs, then only about 10 percent of the eggs on average would be eaten. Dilution is as reasonable a working hypothesis as mate copying, and only direct experimentation (not yet undertaken) can tease them apart.

It may turn out that all the alternative reasons that females choose eggs are wrong and that female mate copying really is what is truly going on in these systems. But as of now, such a claim would be premature.

One loud and clear message comes through from the work on mate copying in guppies: a small brain is not necessarily a barrier to imitation. This counterintuitive finding turns everything we think about culture on its head. Culture is not just for "higher" animals. Nor is it a sign of intelligence. The plain implication is that it is a much more fundamental force than we have ever thought before.

Steve Shuster and Michael Wade are two excellent evolutionary biologists who have substantially made their careers by showing how important extremely simple creatures are for our understanding of behavioral and evolutionary biology. In an attempt to test the strength of one of the models we discussed in Chapter 2, Shuster and Wade examined female mate choice in a marine bug—more precisely, the marine isopod *Paracerceis sculpta*. In this species, breeding occurs in small sponges defended by males. Shuster and Wade found that "in the field, large breeding aggregations contain a disproportionate number of recently inseminated females, suggesting that large harems are particularly attractive to sexually receptive females." They plugged in the differences in reproductive success of males based on the number of females they had on their sponge. What they found was that for much of the year, the copying parameter in their model was significantly different from random. According to their model, females copied each other's choice of mates.

To truly demonstrate mate copying, Shuster and Wade needed to address the following question: Why are females attracted to sponges that contain other females in this species? They suspect that the answer is that it is dangerous to move from place to place in the isopod environment and that these bugs appear to regard the odor of other females in sponges as a sign of a good-quality breeding site. So females are attracted to other females, but what is not clear is whether females are using *the mate choice of others* as a factor in their own decision-making process. Female isopods do not seem to be observing the choice of others in the common sense of the word. So until further work is done in this system, it might best be thought of as an example of "female aggregation" rather than female mate-choice copying.

The deer, stickleback, and isopod work demonstrate that sometimes if it walks like a duck and talks like a duck it still may not be a duck, if you are careful and conservative in your definition of what constitutes a duck. For the scientific investigation of the cultural transmission of behavior in animals (including humans) to make progress, clear, testable definitions of what does and doesn't measure up are needed. As we saw, despite first appearances, systems that appear to include mate copying sometimes do not. In many ways, that is good. It needs to be clear that cultural transmission is not weakly and fuzzily defined as encompassing things that clearly don't meet some baseline minimum criteria. The guppies meet this criteria, but they aren't the only ones. In order to get a more complete understanding of the dynamics of cultural transmission and mate choice, we need to take a look at a few other well-studied cases.

LIFE ON LEKS: THE ROLE OF CULTURAL TRANSMISSION

In the introductory chapter, we briefly examined a study by Jacob Höglund and his colleagues in which they used stuffed dummy females placed on male grouse territories. In this experiment, some males had stuffed black grouse dummies placed on their respective territories early in the morning, before females arrived. These males courted the dummy females and even mounted them and attempted numerous copulations. Höglund and his colleagues found that females were more interested in a male with other (dummy) females on his territory. Such a finding is precisely what one would expect if copying, rather than some set of physical traits alone, was at play on black grouse leks.

Copying may also play a prominent role in the mating system of another related bird species, the sage grouse (*Centrocercus urophasianus*). This is another arena or lek breeding species, which means that males defend small areas that typically have few or no resources in them. The quality of a male's territory on a lek then is thought to have minimal effects on whether a female chooses him as a mate. What makes this type of mating system so interesting to behavioral ecologists (especially those interested in cultural transmission) is that in many lek breeding species, a single male on a lek will get up to 80 percent of all mating opportunities. If it isn't a male territory that gets him mates, then there are at least two obvious reasons that mating success is so lopsided on leks. Either males differ in quality and most females independently decide on the same high-quality male, or females copy each other's choice of mates. Naturally, a bit of both is also possible.

Robert Gibson and Jack Bradbury have been studying the complex mating dynamics on sage grouse territories for more than fifteen years. In the early 1990s, in collaboration with Sandra Vehrencamp, they addressed the question of female mate copying in grouse directly for the first time. Gibson and his colleagues examined the mating behavior of sage grouse females in two different leks over a four-year period. One of their hypotheses regarding imitation and mate choice was that the unanimity of female mate choice would increase as more hens mated on a given day, because more opportunities to observe and imitate would exist on such days. Sure enough, the data support this hypothesis. Further, their analysis reveals the snowballing effect sometimes found in conjunction with imitation. Not only did females copy each other, but copiers copied other copiers.

Gibson and his collaborators also found that females, in addition to copying one another's choice, did some assessment of male traits directly, and this too had a significant effect on mate choice. What is perhaps most intriguing about this system is that copying and assessment seem to interact in a way that produces an unexpected but important phenomenon. They found that the traits that females preferred in males differed between leks not separated by great distances. That is, the traits that females found attractive in males on one lek were different from the traits females found attractive in males on another nearby lek.

The most plausible explanation for this finding is that some natural variance exists in exactly what females find attractive. If a few females assess males early on and subsequent females copy their choice, then our initial assessors have a tremendous impact, and any differences among them will be dramatically magnified by copying. Consider the following hypothetical case of birds on two leks, just a few miles from each other. Suppose that in both lek 1 and lek 2, all else equal, 80 percent of the females like type A males and 20 percent like type B males. If females simply assessed males on their own, then in both lek 1 and lek 2, we should find type A males obtaining about 80 percent of the matings. But what if just two females assess males at the start of the process, and most others copy? Then it is very possible that in one lek, our early assessors choose male A, and, by chance, both early assessors choose male B in the other lek. Instantly, layering copying on this system not only dramatically affects what happens at a given location, but explains much of the difference we see between groups at different locations.

There are many species of lekking birds (as well as other animals). The key to searching for copying in such systems seems to lie in the variation in reproductive success among males. When division of reproduction is divided up more evenly or there is a scarcity of other females to observe, copying is unlikely to occur, as in the great snipe and the pied flycatcher. However, as is more often the case, a single male will garner most mating opportunities. Such systems are ideally suited to studying the cultural transmission of mating behavior.

FROM MATE COPYING TO DATE COPYING

After settling in at a new university, one of the first things I do is start exploring other departments in search of compatriot researchers also interested in behavior. Naturally this often lands me in psychology departments, and that is just what happened when I joined the biology department at the University of Louisville. One week I was invited to give a "brown bag, lunch bunch" talk to the psychology department, and I immediately said I would be delighted to do so. Basically I spent an hour showing slides of the guppy mate-copying work. Psychologists have a penchant for studying copying behavior in humans, and I thought that in addition to telling them about some animal work, perhaps I might pick up a few words of wisdom on the general topic of copying.

After my talk, a graduate student named Perri Druen introduced herself and informed me that she was very interested in asking the same questions I asked of guppies, only in humans. I had always fancied doing just that but didn't have the training in working with human subjects to think I could pull it off. When that potential expertise fell in my lap, I knew I could not let the opportunity slip away. Druen's mentor in the psychology department, Michael Cunningham, had an international reputation for studying attractiveness in humans, and Cunningham too was interested in such a study. (It was a sad indication of the compartmentalization of science for me to learn that Cunningham's office was in fact down the hall from mine. Biologists don't tend to venture into the psychology department. This rule of thumb holds true even in my own university, where both departments are in the same building.

Before long Cunningham, Druen, and I joined with another of Cunningham's graduate students, Duane Lundy, and began the first study that we know of that examines what I shall call date copying in humans. Our basic question was whether social information about whether someone else finds a potential date attractive influences your evaluation of such a person. Further, we wanted to know how strong such socially derived information was in comparison to other variables, such as physical attractiveness. I had evaluated such questions in guppies, but guppies can't fill out a form and tell why they choose whom they choose (although one can make some inferences), while humans most certainly can. Finally, a small part deep inside me wanted the high school quarterbacks of the world to know that science was going to show that looks weren't everything, and it was going to be a nerd who grew into a daddy who did it.

We tested seventy-four female undergraduates and sixty male undergraduates at the University of Louisville. Each woman was tested separately and was told that she was part of a general survey on dating habits. Here is the information given to each woman tested:

> Sandy [a male] was interviewed independently by five women in a previous experiment. Each interview lasted 20 to 30 minutes, and the interviewer was allowed to ask anything she wanted. The five women then rated Sandy on several characteristics. These ratings were made using 10-point scales, where the higher the number, the more positive the rating. In terms of physical attractiveness, his average rating was **3** out of 10 (where 1 = extremely physically unattractive and 10 = extremely physically attractive). In addition, the five women were asked to indicate how interested they would be in dating Sandy. **Four** of the five women indicated an interest in dating Sandy.

Numbers in bold were variables that my colleagues and I manipulated. That is, while one group of women was told that Sandy received a certain score for physical attractiveness and "interest in dating" (I shall refer to this as "popularity"), another set of subjects might see the same survey but with different values for attractiveness and popularity. In addition, the essay that subjects read was easily changed to handle male subjects (again, students at the University of Louisville) who were evaluating a potential date with a female interviewee.

After reading the essay, females were then asked to answer the following questions:

1. How interested would you be in dating Sandy?
2. How interested would you be in marrying Sandy?
3. How good do you think Sandy's social skills are?
4. How good do you think Sandy's sense of humor is?
5. How wealthy do you think Sandy is?

I wasn't thrilled with question 2 about being interested in marrying Sandy. It is a silly question, since no one can give a reasonable answer. But this was a standard question that my colleagues in psychology asked on their surveys, and it was always possible that my assessment was wrong. I admit I never mentioned my doubts to my colleagues.

What we predicted was that females would be more affected by popularity than males. Generally females tend to be choosier than males about their mates, and so we reasoned that this additional information on the preferences of others would be a valuable piece of information for choosy individuals. Further, in many studies like this one, males are more concerned with physical attractiveness than are females, and so we expected this in our trials as well. What we in fact found was that males and females were concerned with both popularity and attractiveness. The more popular or attractive someone was (as reflected in the survey), the more our subjects expressed an interest in dating that person, regardless of the sex of the subject. So, while physical attractiveness was important, individuals also copied the choice of others. If Mary said she'd go out with Sandy, that information made Susan more likely to express an interest in dating Sandy. And if Steve was interested in Cindy, Rick was as well.

The most interesting finding in our date-copying experiment was not how important popularity was, but what traits were attributed to popular individuals, and here a real difference between the sexes did emerge. Remember that one reason to do this study was that while the animal work showed a copying effect, it is impossible to know what new attributes an observer female animal bestowed on a male just selected by another female. Our final three survey questions, however, allowed us to ask just that question in people.

We posed three questions: How good do you think X's social skills are? How good do you think X's sense of humor is? How wealthy do you think X is? What we discovered is that while both males and females attributed social skills, a sense of humor, and wealth to popular individuals, a critical difference between the sexes emerged: Our female subjects clearly stated that they assumed if others were interested in a particular male, that male was wealthy. Why they made that leap we can't say, but that they made it is unambiguous.

We can't say exactly why females attributed popularity to wealth, but other studies provide some hints. For example, David Buss ran a cross-cultural survey on human mate choice and predicted (based on evolutionary models) that females should be more concerned with a male's income than vice versa. Buss found that this was the case in thirty-six of the thirty-seven cultures he examined. Perhaps it is not so surprising that females attribute wealth to popular males. Our subjects may well have assumed that other females viewed a male's income as critical, and that is why a particular male was popular.

SONGS OF JOY

The impact of cultural transmission on mating is not always so direct. Although all of the examples we have examined so far focus on dramatic changes in mate choice in very short order, this is not a prerequisite for the cultural transmission of behavior. In humans, many of the things we learn via culture as toddlers and children don't truly come into play until we are much older. So it is with song learning in birds.

Song learning in birds varies across species. In some species, songs are primarily used during potentially aggressive interactions, while in others they are used in courtship. In some cases, a song may contain a single song phrase, and in others it may contain thousands of such phrases. Some species may have critical periods during which learning may take place. Some birds can learn only their own species' song; in others this is not a constraint. But one thing that is common to all songbirds is that they *learn* the songs they sing. In particular, much of the song-learning process involves learning songs from others (often referred to as tutors). Songs, in other words, are culturally transmitted.

Cowbirds might seem an odd species in which to study the cultural transmission of behavior and its consequences for mate choice. Cowbirds are similar to cuckoo birds, in that they always lay their eggs in the nest of *another* species. So in nature, cowbird young are never raised by adult cowbirds. Hence, it has long been thought that much of the behavior shown in this species must be innate. After all, one must "figure out" how to be a cowbird, and if you aren't learning it from mom and dad, then "cowbirdness" must lay almost completely in the genes. However, cowbird songs demonstrate that information can be transmitted early in life and significantly affect the adult's choice of a mate.

The strange thing is that cowbirds learn their songs—and there are many different "subpopulations" of cowbirds, each with its own slightly different song repertoire. Furthermore, individuals from a given population prefer mating with others that sing the song native to their own area. Despite all the conjectures about genetics then, song learning, a cultural variable, seems to play an important role in mate choice in cowbirds.

Todd Freeberg undertook a series of fascinating experiments with cowbirds to get a better understanding of the cultural transmission of birdsong and its long-term consequences for mate choice in this species. He went about collecting birds from two different populations of cowbirds: one from South Dakota and one from Indiana. These populations were chosen because the cowbirds in them are very different in terms of the behaviors they display and the songs they sing; they may even be genetically distinct from one another.

Freeberg ran what is technically referred to as a cross-fostering experiment. He raised juvenile birds from the Indiana population with either Indiana adults or South Dakota adults. Juvenile birds from South Dakota were also raised with adults from either their own or the Indiana population. Cross-fostering experiments are a well-established technique for studying the effect of an animal's environment. If juveniles take on traits of the environment in which they are raised, regardless of whether it is their "natural" environment, this suggests a certain level of behavioral flexibility.

After juveniles were raised with adults for a year, their own mating patterns were observed in a big aviary. All individuals were tested in an aviary that contained *unfamiliar* birds from both the Indiana and South Dakota populations. Freeberg found that birds preferred to mate with individuals that came from the same rearing treatment they themselves came from, even though they had never before interacted with the particular birds in the mating aviary. His results were even more pronounced when the test was repeated after birds had been together longer.

Freeberg's findings show the importance of cultural transmission in shaping mate choice. First, they clearly demonstrate that information conveyed during early life may not have an impact until individuals mature, but the impact, when it occurs, is quite strong. Second, this work suggests that cultural transmission manifested itself in two different ways. Not only did males apparently learn what song to sing (and how to behave) by imitating the actions of the adults they grew up with, but females also appeared to learn which male traits to view as attractive by watching their mentors.

Freeberg, however, was not satisfied with just demonstrating that the models juveniles use affect mate choice as an adult. To make his case rock solid, he examined *the offspring* of birds that obtained their songs via cultural transmission. This second-generation experiment was similar to the first one. Fifty-six juvenile cowbirds were captured at one of the South Dakota sites. These birds were divided into two categories for separate treatments, which I shall call SD/SD/SD and SD/SD/IN. SD/SD/SD birds were housed with adult birds that came from South Dakota, and the adult birds themselves were trained by birds from South Dakota. SD/SD/IN cowbirds were housed with adult birds from South Dakota, but these adults had been trained by Indiana birds.

Once the juvenile birds matured, Freeberg put cowbirds from the two groups together in such a way that members of the opposite sex were completely unfamiliar with each other. Cowbirds preferred to mate with individuals who sang like their tutor's tutor. The strength of the cultural force is clearly significant if who tutored your tutor affects whom you choose. Given that there are many species of songbirds and that all learn most of their repertoire from listening and observing others, song learning in birds is and will continue to be a hotbed of work on the evolution of cultural transmission for many years to come. But this result in itself is already staggering.

A MODEL OF CULTURE?

In precisely what manner culture operates across the animal kingdom remains to be answered, but the evidence plainly shows that culture is a force to be reckoned with in nature. We need to rethink our views on how this force operates. To begin with, despite early thinking on the subject, cultural transmission is anything but limited to the most cognitively sophisticated animals (primates, for example). In fact, most studies on this phenomenon have been undertaken in what used to be referred to as "lower" vertebrates. From guppies to birds, animals with relatively small brains assimilate some sort of cultural rules into their mating behavior. Cultural transmission, then, is not correlated with brain size.

That is big news indeed, for it demonstrates not only that humans do not have a monopoly on a powerful evolutionary force, but neither do primates or any group of charismatic megafauna.

Of course, the species bias toward studying culture and mate choice in lower vertebrates is likely due to the speed and control typically associated with nonprimate studies, and no doubt once studies on cultural transmission and mating are undertaken in primates, even greater advances in our understanding of behavior will follow.

With a firm grasp now on how cultural transmission affects mating in animals, we are ready to move on to perhaps a more basic question: Why do we see cultural evolution in the first place?

4

The Meaning of Culture

I am a firm believer that without speculation there is no good and original observation.

CHARLES ROBERT DARWIN TO ALFRED RUSSELL WALLACE, 1867

DARWIN WAS RIGHT ABOUT THE RELATIONSHIP between speculation and observation. Without a hypothetical structure to work from, empirical studies are an arbitrary and meaningless accumulation of facts.

Developing a theory is part and parcel of all scientific endeavors; however, it is particularly important with respect to evolution and culture. To begin with, the examples in the previous chapter, however jazzy they were, are not the ultimate answer to the mysteries of evolution, culture, and the more practical problem of choosing a mate. The results of the studies we examined, intriguing as they may be, are in a sense just pieces of information. The scientific process is fundamentally about generating models of the world and putting those models to the test. Up to this point, we haven't done much theory generating.

Culture means many things to many people. Robert Boyd and Peter Richerson have noted that even back in 1952, A. L. Kroeber and Clyde Kluckhohn identified 164 definitions of culture. No doubt the list is longer today. Researchers in anthropology, psychology, biology, sociology, political science, and economics study culture in one way or another, and most people view *culture* as such an amorphous term that it can't be studied scientifically. The importance of a solid definition for the purposes of our discussion cannot be put off any longer. Following Boyd and Richerson, culture "is information capable of affecting individuals' phenotypes which they acquire from other conspecifics by teaching or imitation," where "phenotype" is a composite of the traits an individual possesses.

If we are to understand what culture denotes in the context of evolution and behavior, we need to have a well-defined theoretical base so that the scope of our predictions can be understood. Mathematical models provide the tools to do just that. Yet models need not be strictly mathematical, but can instead be conceptual. Our progression here will be from the more conceptual to the more mathematical.

Let's begin by being specific about our terms. While we have already used the phrase *cultural transmission* on a few occasions, it is now time to be more specific about this key concept. Cultural transmission refers to the set of ways that culture can spread through a population. For example, one can imitate or teach in various different ways to various different types of individuals. Cultural evolution, on the other hand, refers to the summed effect of cultural transmission over long periods of time.

With these definitions in hand, we can now move on to the theoretical work underlying the study of evolution and culture. The "theoretical pie," as we shall see, can be divided many ways.

THE ROLE OF GENES IN CULTURAL TRANSMISSION

I have presented seminars on mate copying at more than twenty universities around the world, and at every single one of these talks, someone has stood up and said something akin to the following: "Okay, I buy that guppies are copying each other's choice of mates, and I will even let you call it cultural transmission if that makes you happy. But isn't it possible that a guppy's tendency to copy is controlled by a gene?" The answer is that it certainly is possible, and we are in fact running experiments on this very question in my lab right now. More important, this question raises two deeper issues. First, suppose that the tendencies of guppies to copy were controlled by a gene. In fact, suppose that this was true for every case of imitation ever recorded. Would this mean that cultural transmission was really just the long hand of the selfish gene at work? If so, why even speak of culture in animals? Second, is it even possible in principle to imagine how cultural transmission in animals would work if it weren't under some type of direct genetic control?

Approximately 80 percent of the female guppies my colleagues and I have tested in experiments copy the mate choice of others. But suppose for the sake of argument that 100 percent of the fish used a mate-copying strategy. Further, suppose that I found a geneticist to work with me, we actually located the gene(s), and then a molecular biologist colleague of ours sequenced it. The fact that 100 percent of the guppies tested had a gene for copying tells us that copying has a genetic basis. But that is all that piece of information tells us, and it isn't enough. If we assume that copying is a good strategy (and that's why the gene is there to begin with), then we actually learn little about how female mate choice will *change* over time. To begin with, females can copy only when given the opportunity to observe others, and in nature that opportunity will be there at times and absent at others. This suggests that we need to know the frequency with which females rely on copying to choose a mate in order to understand how male and female sexual traits will change over time. Thus, we can see that genetics can't explain cultural transmission as an evolutionary force.

Whom female copiers choose as mates will depend on whom they see their "models" choose. This, in turn, depends on whether the models had the chance to copy, and if they did, whom *they* observed. Did that individual copy or not? And on backward infinitum. None of these things can be determined by understanding that 100 percent of the females have a gene for copying. But the dynamics of this system can be more easily understood if one views cultural transmission as a system that involves the transfer of information through social learning. So it would be great to know as much as possible about the genetics of copying, but that wouldn't obviate the need to speak of cultural transmission as a force in its own right.

The second question that arises from the persistent hecklers that come to my talks is whether it is even possible in theory to have cultural transmission without some underlying genetic component. That is, is it possible even to imagine a system of cultural transmission that is completely divorced from any genetic underpinnings? A system that perpetuates even in the absence of a genetic component to imitation?

In short, yes. The first type of cultural transmission that can exist in the absence of any genes coding directly (or even indirectly) for it is teaching. If individuals teach others how to undertake certain activities, and then individuals who learn teach others, you have a self-perpetuating system of cultural transmission completely divorced from genes. Of course, it is possible that the tendency to teach may have a genetic component, but the point here is that it is not necessary for this to be part of the system in order for it to work. If individuals start teaching each other for any reason and those who learn then teach others, the whole system takes off.

Direct teaching is not a necessary part of a system of cultural transmission independent of genetics. Imagine a case where some individuals start copying the actions of others (no teaching here, just copying). Further, imagine that individuals are able to assess when copying the actions of others pays better than not copying, and cultural transmission is again off to the races. This system incorporates both social learning (copying) and individual learning (learning the costs and benefits of various options open to you), and it works especially well if individuals copy the most successful (or even the most common) strategy that others use. Boyd and Richerson refer to examples like this as "biased transmission," because the particular behaviors that are copied are biased toward those that are most successful.

Unfortunately, the state of empirical work on imitation and mate choice is not sufficient to determine the frequency of the two types of cultural transmission described above (one based on a gene underlying the tendency to copy and one independent of genes). We are not able even to classify a case of mate choice and cultural transmission as being purely driven by a genetic tendency to imitate or as being completely separated from genetic influence. The sort of detailed experimental work on the costs, benefits, and heritability of imitation needed to distinguish between these types of culture has simply not been undertaken in any system.

WHEN DOES CULTURAL TRANSMISSION PAY?

When you get down to nuts and bolts, most of what goes on in life centers around getting information of one type or another. And there are only three ways to possess information: you could always have possessed it, you could have learned it yourself, or you might have obtained it from others. In terms of science, these three paths to information acquisition translate into genetic coding, individual learning, and culture. Much of what some animals "know" unquestionably comes directly from genetic programming. Genes code for certain behaviors, and with respect to these behaviors, animals "know" what to do instinctively. The question is: Is cultural transmission a better means for accumulating information than genetic reproduction? It depends.

Theoreticians begin by simplifying the problem. Let us examine two of the three paths to acquiring information: genes and individual learning. Once we understand something about what favors each of these when they compete against each other, we can stir social learning back into the stew.

At the most basic level, behavioral ecologists and psychologists have long said that learning is favored over genetic transmission when the environment an animal lives in changes frequently, but not too frequently. The logic here begins with the assumption that there is some cost to learning, even a very small cost. Then, when the environment never changes, information is best passed on genetically, to avoid the cost of learning. This works because the environment of mom and dad is similar to that of junior. If the environment is always changing, there is nothing worth learning because what is learned is completely irrelevant the next time interval down the road. So again, genetic transmission is favored. Somewhere in the middle, in between an environment that never changes and one that always does, learning is favored over genetic transmission. In this zone, it is worth paying the cost of learning. Here, the environment is stable enough to favor learning, but not so stable as to favor genetic transmission.

That primarily verbal argument has been challenged by David Stephens as being too simplistic. Stephens agrees that the starting point in modeling the fight between genetic transmission and learning should indeed be the stability of an animal's environment. But, Stephens argues, environmental stability as it is represented in these verbal models mixes up two types of stability that need to be separated.

Rather than think of environmental predictability as a single force, in Stephens's model predictability is broken down into two parts: predictability *within* the lifetime of an individual and predictability *between* the environment of parents and offspring. These two types of predictability can be very different, and lumping them together may hinder our understanding of the evolution of learning. Imagine that early in life, offspring of some species disperse to environments that are far removed from those of their parents.

Further suppose that during the adult lifetime of such individuals, their environment will be relatively stable and thus predictable, even though the environment is unpredictable across generations of this species.

To address when learning is favored over genetic transmission, Stephens constructs a model that is not for the mathematically faint of heart. In so doing he creates equations that capture the cost of learning, as well as the different types of stability, and he adds a few other parameters to boot. Fortunately, the core of his results can be seen in the table shown here. Mind you, this table simplifies many of Stephens's findings, but it captures their essence. Learning is favored in Stephens's model when predictability within the lifetime of an individual is high but environmental predictability between generations is low.

Naturally the model becomes more complicated when more variables are added, but Stephens's simple table cap-

STEPHENS'S EVOLUTION OF LEARNING MODEL

Between-Generation Predictability	Within-Lifetime Predictability: LOW	Within-Lifetime Predictability: HIGH
LOW	IGNORE EXPERIENCE	LEARN
HIGH	IGNORE EXPERIENCE	IGNORE EXPERIENCE

Source: D. Stephens, 1991, Change, regularity, and value in the evolution of learning, *Behavioral Ecology* 2:77–89.

tures the heart of an otherwise complex model. In a nutshell, Stephens shows that when things are relatively constant within the lifetime of an individual, it pays to use what you have learned in the past, because the past is a good guide for the present. But this sort of predictability is usually what favors genes, right? Sure, but the reason genetic coding fails here compared to learning is that the world changes dramatically *between* generations, and what worked for mom and dad wouldn't necessarily work for you, and thus you are better off learning.

So how does social learning, or cultural transmission, affect this model? Boyd and Richerson introduced the notion of "guided variation" to address this question. Guided variation brings individual and social learning together into a single model of information acquisition. In a system of guided variation, individuals observe the actions of others and modify their behavior based on what they have observed. On top of this, learning is invoked to choose among various options that observers have seen. For example, imagine that individuals learn three new foraging techniques by observing the actions of the elder foragers in their group. This, in and of itself, constitutes social learning, but it becomes "guided variation" when observers then use individual learning to determine which of these three newly acquired foraging techniques they will, in the end, adopt. They may, for example, dabble a bit in each new food-gathering mode, and then see which one produced the greatest amount of nutrients at the end of some prescribed testing period.

In this model, Boyd and Richerson are concerned with the relative importance of individual learning and cultural transmission, not just how these factors fare against genetic transmission. Genes, however, do play a role in the model in that it is assumed that an individual's reliance on learning versus cultural transmission has an underlying genetic component. They do consider the case when this is not true, but not in the basic model we are examining. Individuals in this model have genetic parents and cultural parents (the individuals they copy). The question is, When does reliance on cultural parents outweigh the strengths of individual learning?

As with the case of Stephens's model, Boyd and Richerson's mathematics can become overwhelming. But as in Stephens's case, the results can also be boiled down. Cultural transmission is favored in relatively stable environments when the error rate associated with it is lower than the error rate tacked onto individual learning. Error rate, then, turns out to be the key to what we see at the end of the day. If you make more mistakes when learning on your own how to achieve some "goal" than you do by learning from others, then you should rely more strongly on learning from others. Otherwise you should rely on your own ability to learn. For example, imagine that we are concerned with foraging tactics. If the individuals I copy have already culled away some bad ideas with respect to getting food, copying them may save me the trouble of learning on my own what is a good way to get food and what isn't. Under this scenario, if I normally make lots of mistakes in my personal learning process, I may be better off copying others.

Boyd and Richerson's models are quite general and apply to any sort of learning and social learning scenario in which a human or nonhuman may find itself. What, however, can we say about the costs, benefits, and error rates associated with cultural transmission and mate choice? Why copy the mate choice of others when you can choose on your own, through either genetic predispositions or individual learning?

Robert Gibson and Jacob Höglund have suggested two possible reasons that cultural transmission may be favored in some mate-choice scenarios. The first focuses on "discrimination." The idea here is that choosing a mate among a bevy of suitors is no easy task. You need to examine all sorts of traits about potential mates, and it might not hurt to get the opinion of others. If, the argument goes, you are going to have to make a tough choice, why not have as many weapons in your arsenal as possible? Perhaps the choice of others can help you fine-tune your own sense of who makes a good mate, providing you with finer discriminatory skills. Perhaps not. As with the more general models of Boyd and Richerson, it is always possible that by copying the mate choice of others you are receiving bad information. It may be that the individuals you copied know less than you do.

Another putative benefit of mate copying compared to going it on your own revolves around "opportunity costs." Opportunity costs, introduced by economists long ago, reflect the mundane truth that by doing one activity, you are giving up the chance to do something else. There is always a cost to doing A, and that cost is that you could have been doing B, C, D, and so on. In the context of mate choice, instead of assessing a male for his suitability as a mate, females could have been eating, looking out for predators, resting, and so on. If mate copying allows you to get information quickly and frees up your time for these sorts of activities, it should increase in frequency. Once again, however, such benefits must be weighed against the potential of getting incorrect information via mate copying.

Whether discrimination abilities or opportunity costs or some other as-yet-unknown benefit drives mate copying is an empirical issue. Unfortunately, there are few hard data that speak to which, if either, idea put forth by Gibson and Höglund is more strongly supported by the evidence.

MATHEMATICAL MODELS: DIFFERENT TOOLS FOR DIFFERENT PROBLEMS

Just as carpenters have many tools in their toolbox, so too do behavioral ecologists building models of mate-choice copying have an array of mathematical tools at their disposal. Some mathematical tools are designed for tackling very complex problems. Others may not handle the supercomputer issues well, but are built to detect small but important differences between things. There are theoreticians who are trained for years to use one specific mathematical technique. There are those who have never received formal mathematical training and would be thrown out of mathematics meetings as impostors, and yet know enough about many different mathematical tools to address important questions in evolution.

As with many other complex issues, it is impossible to begin with a complete, all-inclusive model of cultural evolution and mate choice. Rather, models are constructed to address specific questions in the context of how culture affects mate choice. There are three areas of investigation ripe for modeling:

1. What impact does female mate copying have on male reproductive success? Do males do better or worse if females copy? Do some do better and others end up alone more often?
2. When might we expect to see females adopt a strategy of imitating the mate choice of others? Would such a strategy always work? Would it depend on how many individuals were copying?
3. How will female mate copying affect the coevolution of innate female preferences for males and the male traits that are preferred by choosy females? We have seen in earlier chapters that sometimes male and female innate traits evolve in unison when it comes to mate choice. What happens when imitation is added to this picture?

We will examine how three mathematical techniques shed light on all these questions of culture and mate choice.

Opportunity for Selection

"Opportunity for selection" models examine female mate copying by exploring what effect imitation on the part of a *female* has on male reproductive success. The underlying idea is that female mate copying probably won't change how many times males in a group will mate, but it will surely change which males get to mate, and how many times. To examine such effects, Michael Wade and Stephen Pruett-Jones used what are called "opportunity for selection" models. In their modeling efforts they were especially interested in examining lek-breeding species. These species are not only full of variance in male mating success, but they are also the most likely to have females employing a mate-choice copying rule because of the abundant chance to observe the preferences of others. Given this, Wade and Pruett-Jones set out to address whether mate copying per se could explain the variance in male reproductive success that is associated with lek breeders.

Wade and Pruett-Jones examined a population in which females breed sequentially and a male's probability of mating with a female is a function of how many females had selected him as a mate in the past. Their results were clear: "Female copying increases the frequency of extreme values in the mating distribution. Whenever there is female copying there are more males that do not mate and more males that obtain large numbers of matings."

One of the strengths of this model is that it can be used to figure out whether mate copying is prevalent in a given population. That is, not only does the model show that mate copying *can* explain variance in male mating success in theory, but the model is built so that data from natural populations can be handily plugged in to determine the extent of mate copying in the system being studied.

Game Theory

Game theory models of animal culture look at the evolution of behavioral strategies when the fitness of one individual is affected by the actions of others. One variant of game theory models entails creating a computer simulation that mimics a real population of males and females adopting different behavioral options. For example, we can look at the success of females that copy the mate choice of others versus the success of females that do not, and then make predictions about whether we should see mate copying in nature. These models make predictions such as "Copying should fare best when females are able to correctly identify high-quality males" and "Younger individuals should copy more often than older individuals." The latter prediction may provide some insight into fascinating human phenomena such as the obsession of teenagers with what older, "in vogue" people are doing.

George Losey and his colleagues built a computer simulation that examined copying and choosing strategies in an environment that simulated a lek-breeding area. Females using the copying strategy were allowed a fixed number of "peeks" at the choices made by other females (copiers and choosers alike). If a copier had not observed any males mate, she either selected just as a chooser would (the "smart" copier model) or she selected males randomly (the "dumb" copier model). However, if she had seen males mate, she chose the male that she had seen mate the greatest number of times. Losey and his colleagues found a mixture of copiers and choosers (with copiers in the minority) to be the optimal solution to the mate-copying game.

Sushil Bikhchandani, David Hirsheifer, and Ivo Welch recently extended the game theory approach in what they call "information cascade" theory. Individuals in these models choose to imitate or not based on the decisions of a series of predecessors. These models, though designed to explain human information transfer, are equally applicable to animals (as the authors themselves note). The power of these newly developed models lies in the fact that rather than just explaining why a particular trait spreads by cultural transmission, Bikhchandani and his colleagues show how fragile cultural transmission can be. Under cultural transmission, what is popular today may be taboo tomorrow.

Fads (a result of cultural transmission) come and go at the drop of a hat. Information cascade theory explains this fragility by demonstrating how small changes in behavior can quickly affect what is acceptable in a society and what is not. For example, information cascades can be used to model drug use, fashion, the stock market, and many other facets of everyday human life. Future work in this particular area of modeling may prove very useful in understanding how cultural transmission can snowball certain behaviors through communities at rates that are magnitudes of order above what genetic transmission can handle.

Population Genetics

One shortcoming of the models already described is that they fail to examine how mate copying affects males and females *simultaneously* (technically called coevolution). In this sense, they miss the forest for the trees. Population genetic models of mate copying, however, tackle the coevolution question head on.

Mark Kirkpatrick and I developed population genetic models of mate copying in which female preferences were affected by both an innate preference and a culturally transmitted preference. In one of our models, immature females observed many adults make choices among potential mates. When these females matured, they had a stronger preference for the male type they most frequently saw mating. Results indicate that culturally determined female preferences and the male trait that copiers end up preferring can coevolve.

Male and female traits in our models "run away" to extreme values when mate copying is allowed. This is analogous to the process of genetic runaway selection we touched on in an earlier chapter, but now the male and female traits become extreme not because of genetics, but because they are linked together by cultural transmission. The reason to take note of this prediction is that runaway genetic evolution has spurred a wave of experiments to see whether one could experimentally demonstrate that male and female genes are linked and consequently move toward extreme values in unison. In practice, before my model with Kirkpatrick, if investigators found such a linked system, they could claim runaway genetic selection without finding the genes involved. This is because there was no other model that predicted a linkage between male and female evolution. Now that cultural runaway models exist, finding that male and female traits coevolve is not enough to proclaim evidence of runaway genetic selection; such a finding could just as easily be the result of runaway cultural selection. So these days, you'd better have the genes located if you want credit for finding runaway genetic selection.

ALL ROADS LEAD TO ROME

Imagine that you are on a neighborhood committee whose goal is to get the children living in your community to abide by rule X. There are obviously many different approaches you could adopt to get kids to live by this rule, but let's focus now on *whom* to use as the conveyers of your rule. Who are the models the kids should learn from? The committee might suggest three different means of conveying the relevant information to kids. First, one time-honored tradition for getting kids to learn is to have their parents teach them, and your committee would be wise to at least entertain this option. Parents have lots of time with their children, and they have the power, at least in theory, to force their children to listen and act in a prescribed manner. Of course, it is often painfully clear to adults that the last person that their children want any advice from is them.

A second popular technique to get children to do something is to have some adult who is not their parent teach them. This is usually accomplished in modern society by arranging for some hero, sports star, celebrity, or other icon to talk to children about the issue of interest. Television, for example, is full of commercials that use icons to convince kids to act in a certain manner. Unfortunately, at least from a parental perspective, such television advertisements are geared toward having children do something their parents are not at all thrilled about, such as spending hundreds of dollars on a pair of improbable looking sneakers. More and more, however, we are seeing civic-minded commercials. Sports heroes are going on the air to convince children to act in ways that tend to please parents: don't do drugs, stay in school, consider abstaining from premarital sex, and so on. The neighborhood committee needs to know that a word from basketball legend Michael Jordan in a child's ear goes a long way.

Finally, there is the most parentally dreaded means of conveying information to children: through peers. If you want children to act in a certain manner, they are much more likely to do so if they are taught by their peers that this is an acceptable mode of action. Children learn from their peers all the time, so if your committee could figure out a way to tap into this transmission mode, they might get their point across in no time. That being said, the information that you are trying to disseminate is much more likely to get corrupted when kids are passing it on to other kids. Remember the game of telephone you used to play as a child. A message starting off as "Aaron doesn't like the rain" may end that way, but it is just as likely to mutate to "Karen likes to play with trains."

These means of transmitting information have specific names in the literature on evolution and culture. The three we have looked at above are labeled *vertical, oblique,* and *horizontal transmission,* respectively. The term *vertical transmission* refers to the case where information of some sort is passed directly from parent(s) to offspring. This could take place any number of ways. Offspring might observe their parents and copy what they do. The finches that Peter Grant and Rosemary Grant have been studying use a form of vertical transmission in both males and females. The males learn the song that they will sing from their fathers, and females learn what songs finches are supposed to sing by listening to their dads. A more sophisticated type of vertical transmission is teaching (we shall return to this subject in some depth later), wherein the parent behaves in a manner suggesting an active attempt at passing down information (copying is more passive than teaching). We obviously rely on teaching a great deal. The animal evidence for teaching is, however, more equivocal, and there are no studies involving teaching and mate choice in nonhumans.

Evolutionary biologists have coined the phrase *oblique transmission* to describe the transfer of information across generations, but not by parent-child interactions. Young simply get their information from model adults that don't happen to be their parents. This sort of transmission might be particularly common in animal species where there is no parental care, in, say, fish, and hence most young-old interactions will be between nonrelatives. Female guppies give birth to offspring, and that ends the mother-child relationship. Fathers have been long gone by this time. Young guppies interact with their elders, just not their parents.

Cultural transmission works within generations, as well as between them. Information need not be transferred from adult to child. Everyday experiences demonstrate that most of the information we get in fact comes from peers—people in our own approximate age group. This phenomenon, called *horizontal transmission,* holds true not only for adults but for children as well. In fact, horizontal transmission of information is so powerful that parents spend much of their time trying to subdue its effects on their young. We want children to learn some things from their peers, but not everything.

Horizontal cultural transmission plays a large role in non-human animals as well. Many of the examples we walked through in the previous chapter were driven primarily through horizontal transmission.

It would be ideal if I could end this section by saying something to the effect that "one can now examine the different predictions that horizontal, vertical, and oblique models of cultural transmission make with respect to mate choice." Unfortunately, we are not yet at that point. In truth, we cannot make this statement for any behavior, let alone mate choice. As studies of cultural transmission and the evolution of social behavior continue, however, we may someday be able to make just such proclamations.

Now that we have seen how models of culture work to enhance survival, the question is, Does cultural transmission of information ever decrease our chances of survival; or as biologists say, Is it ever maladaptive?

BAD CULTURE?

Consider this rather odd story about yams:

> According to William Bascom (1948), on the Micronesian island of Ponape a man's prestige is partially determined by his ability to contribute very large yams to periodic feasts. Each year several feasts are given by the chief of each district. In addition to sta-

> ple foods like breadfruit, coconut and seafood, the head of each farmstead contributes a "prize" yam or yams. Everyone at the feast examines the yams and praises the contributor of the largest yam for his generosity and his skill as a farmer. Moreover, as Bascom reports: "Success in prestige competition is regarded as evidence not only of a man's ability, industry and generosity, but also of his love and respect for superiors. The chiefs raise men who are consistent contributors of large yams to titled positions. . . . Moreover, these yams are truly huge; they sometimes exceed 9 feet in length and 3 feet in diameter, and up to twelve men must carry them."

The story gets even stranger in that it doesn't matter what else a farmer brings to the feast, and it doesn't matter how many yams he brings. All that matters is the size of the biggest yam.

How could such a bizarre ritual focusing on yams have evolved? It's not as if there are no shortages of food on the island, so the opportunity to muck about with yam size for a leisure-time activity can't explain what we see. Food shortages are common on this island, and families have been known to starve while growing a huge yam for festival time.

Is it theoretically possible that cultural transmission can lead to behaviors that are *genetically* maladaptive, that is, behaviors that would normally be slowly culled by natural selection? Yes, Robert Boyd and Peter Richerson argue; "runaway" cultural selection can explain the strange story of yams and the people of Ponape. The argument goes like this:

> Suppose that at some earlier time Ponapeans did not devote any special effort to growing large yams. It seems reasonable that under such conditions more skillful or industrious farmers might have tended to bring larger yams to feasts, and thus the size of a man's yams would provide a useful indicator trait for all kinds of skills and beliefs associated with farming. By imitating the people who grew large yams, naive individuals could increase the chance that they would acquire the cultural variant they needed to be successful farmers. Once the size of the yams became an indicator trait, beliefs or practices that lead to larger yams would increase. Individuals with a stronger tendency to admire large yams will be more likely to acquire these beliefs. This will cause the two traits to be correlated—and therefore, when the practices that lead to larger yams increase, so too will the admiration for the ability to grow large yams.

The next thing you know, you have nine-foot yams and full-blown yam heroes.

Don't think that this kind of behavior is limited to humans. On the nonhuman side of the animal kingdom, it turns out that the social transmission of information can, under the right (or more appropriately, wrong) circumstances, lead guppies to undertake some seemingly very maladaptive foraging behaviors.

To get a more comprehensive understanding of how cultural selection might produce maladaptive behavior, consider the most blatant case for a culturally derived behavior: reproduction in modern humans. Here we see yet another mechanism by which cultural evolution can produce genetically maladaptive behavior.

Even the most basic understanding of the Darwinian process of natural selection would suggest that the number of offspring per parent should be relatively high in modern Western societies (especially industrialized countries). After all, natural selection should optimize the number of children people are having, and that means that the more resources available in a given *population,* the more offspring that population should be producing. Furthermore, the greater the resources available to a particular *family in a population,* the more that family should be pitching in to the next generation's gene count. Yet the exact opposite appears to be the case in modern Western societies. Birthrates are lowest in industrialized nations, and within such nations, the richest families have the fewest children. What's going on?

To explain this reproduction dilemma, Boyd and Richerson introduce the concept of "asymmetrically transmitted variation." This jargony phrase is translated by its coiners as follows:

> Individuals characterized by a particular cultural variant may be more or less likely to achieve a given social role than individuals with other cultural variants. This means that the attainment of many different social roles entails surviving a selection process. The variant that maximizes the probability of attaining social roles other than that of parent may often be different from the variant that maximizes the probability of becoming a genetic parent. If individuals occupying social roles other than parents are involved in cultural transmission, then many of the selection processes that act on cultural variation may increase the frequency of genetically maladaptive cultural variants.

In the case of declining reproduction, Boyd and Richerson argue that attainment of the most prestigious and sought-after sociocultural roles is likely to require traits that are in conflict with pure genetic reproduction. Becoming a professional (arguably a socially desirable position) requires time and money. If you want your children to get there, the argument goes, you'd better not divide up your resource pie among too many kids. Boyd and Richerson even argue that there is some indirect evidence that being in large families does hinder attaining professional status. For example, professional status often (but not always) requires a certain degree of intelligence. If one assumes that IQ is a measure of intelligence, then the evidence that large families produce children with lower IQ scores bodes well for Boyd and Richerson's hypothesis.

Exactly how many kids we expect people to have on average depends not on a simple selfish gene head count, but on the relative strength of cultural processes in relation to classic natural selection processes. In the case outlined above, the stronger the cultural forces are, the fewer the children produced in resource-rich societies.

In our discussion of models of cultural evolution, there is one popular approach that we have failed to touch on as of yet, and that is the idea of "memes" and cultural transmission. A "meme-based" view of much of human behavior has received great attention as of late, and merits its own chapter. It is to this topic that we now turn.

5

Meme Again

Da-da-da-daaah, da-da-da-daaah . . .

Opening of Beethoven's Symphony no. 5

MY *WEBSTER'S* DICTIONARY DEFINES *GENE* AS "a functional hereditary unit that occupies a fixed location on a chromosome." Now there is something you can sink your teeth into. No matter how genes work, people have a certain confidence in them because they know they have a fixed location on one of those curly little things known as chromosomes. We talk of a trait as "being in the genes," and, although for purists who spend their life immersed in genetics the *Webster's* definition is somewhat lacking, for everyone else it works just fine, thank you—so fine, in fact, that a sort of "gene envy" exists and has long existed among those interested in culture and cultural transmission.

Wouldn't it be great, the pundit of cultural transmission

muses, if we had a word that carried the same punch as *gene*? Something that captured the essence of cultural transmission in a single, short, but catchy moniker? Perhaps people then would take this cultural transmission stuff seriously.

Given the musings of our hypothetical researcher, it should not come as a shock that many people have attempted to coin a phrase to serve as the cultural equivalent of a gene. The idea has always been to try to create the analogue of *gene*—something that was discrete and somehow passed down through the generations yet was distinctly cultural instead of genetic. Contenders for this title include Charles Lumsden and E. O. Wilson's *cultragen,* Cloak's *i-cultures,* and L. L. Cavelli-Sforza and Marcus Feldman's *cultural trait.* The undisputed champion, however, is the concept of the *meme.*

In the last chapter of his classic *The Selfish Gene,* Dawkins, champion analogy coiner, introduces the reader to the meme:

> We need a name for the new replicator, a noun which conveys the idea of a unit of cultural transmission, or a unit of imitation. "Mimeme" comes from a suitable Greek root, but I want a monosyllable that sounds a bit like "gene." I hope my classicist friends will forgive me if I abbreviate mimeme to meme.

Definitions here are critical, so it is important to note that there have been many attempts at deconstructing or decomposing memes, and the process of defining them has itself been an evolutionary one. Let's take a look at a few attempts at reinventing the meme concept:

> An element of a culture that may be considered to be passed on by a non-genetic means, esp. imitation.

> A contagious information pattern that replicates by parasitically infecting human minds and altering their behavior, causing them to propagate the pattern. (Term coined by Dawkins, by analogy with "gene.") Individual slogans, catch-phrases, melodies, icons, inventions, and fashions are typical memes. An idea or information pattern is not a meme until it causes someone to replicate it, to repeat it to someone else. All transmitted knowledge is memetic.

> A meme should be regarded as a unit of information residing in a brain. It has a definite structure, realized in whatever physical medium the brain uses for storing information.

> A meme is whatever it is that is passed on by imitation.

> A unit of cultural inheritance. Hypothesized as analogous to the particulate gene, and as naturally selected by virtue of its "phenotypic" consequences on its own survival and replication in the cultural environment.

I find the last of these definitions the most satisfying, both for its content and because it is the one Dawkins proposed after really thinking about memes for a while.

At the start, Dawkins's meme concept was viewed as intriguing, but it never really caught on, even among evolutionary biologists. But that has changed. Over the past five years, the meme idea has proven to be a successful meme itself, and the new field of memetics has arisen, with its own on-line journal and popular books. To followers of this new subdiscipline, studying memes is the key to understanding the evolution of culture. To solidify the field and give it the appearance of a serious discipline, a new lexicon of mimetic terms has been created. As a sample, here is a section from the memetic lexicon Web site:

> **co-meme** A meme which has symbiotically co-evolved with other memes, to form a mutually assisting meme-complex. Also called a symmeme.
>
> **membot** A person whose entire life has become subordinated to the propagation of a meme, robotically and at any opportunity. (Such as many Jehovah's Witnesses, Krishnas, and Scientologists.) Due to internal competition, the most vocal and extreme membots tend to rise to the top of their sociotype's hierarchy. A self-destructive membot is a memeoid.

meme-complex A set of mutually-assisting memes which have co-evolved a symbiotic relationship. Religious and political dogmas, social movements, artistic styles, traditions and customs, chain letters, paradigms, languages, etc. are meme-complexes. Also called an m-plex, or scheme.

memotype 1. The actual information-content of a meme, as distinct from its sociotype. 2. A class of similar memes.

meta-meme Any meme about memes (such as: "tolerance," "metaphor").

What makes memes so very special in the eyes of Dawkins and other concerned citizens is that they have achieved a status that until now has been reserved for genes. Memes are replicators, and from an evolutionary perspective, the argument goes, that makes all the difference. A replicator has three properties: fidelity (good copies are made), fecundity (lots of copies are made), and longevity (copies are made for a long time). Dawkins sums up replicators nicely as "any unit of which copies are made, with occasional errors, and with some influence or power over their probability of replication."

To bring the idea of a meme closer to home, consider the birthday parties that are a ritual in most Western societies. When was the last time you went to such a party and failed to hear everyone sing the birthday song? My guess is that it was a long time ago (if ever). And, of course, there is little variation in what song people sing: "Happy Birthday to You." There is no other song for the occasion. The tune is a meme, and a successful one.

In order to see why "Happy Birthday to You" is a meme, recall the three characteristics that define a replicator: fidelity, fecundity, and longevity. You don't hear many people, even little children, making mistakes when singing the song. When the rare mistake is made, it is inevitably by a single novice singer who is immediately cued back in to the correct tune by the rest of the singers. Further, when young children are taught "Happy Birthday to You," they are all taught virtually the exact same song. Fidelity (how accurately a replicator is copied) is not a problem. Add on to this the fact that virtually all children learn "Happy Birthday to You" early in life, and that this has been the case for many years, and you see that "Happy Birthday to You" meets the fecundity and longevity criteria of a replicator as well.

One might argue that while "Happy Birthday to You" meets the criteria of fidelity and fecundity, longevity is more of a stretch. After all, this tune has been around for only a couple of generations, a drop in the hat in evolutionary terms. The trouble is that it's not exactly clear what constitutes longevity in a replicator. In any case, recall that Beethoven's Symphony no. 5 and "In the beginning G-d created the Heavens and the Earth" are also good candidates for memes, and they have been around for hundreds and thousands of years, respectively. "Happy Birthday to You" might not leave the party for a significant period of time, even by evolutionary standards.

What makes replicators like memes unique is that they are the "beans" of the evolutionary accounting system, and replicators have only one "objective," and that is to make more copies of themselves. They are not here for anything else, just to make copies of themselves. The entire *Selfish Gene* is really an expanded discussion of why genes are replicators. And until the concept of a meme popped up, genes were believed to be the only replicators on earth. Sure, other replicators were possible in principle, but no one had staked a strong claim for one, until Dawkins raised the possibility that memes are replicators in the same sense that genes are. A solid meme has the same three characteristics—fidelity, fecundity, and longevity—and that makes it a replicator, in just the same way that a gene is a replicator.

Once we recognize the meme as a replicator, we stop thinking once and for all that cultural traits are always, one way or another, selected because of the effect they have on genetic fitness. If memes are replicators, they are interested in making more copies of memes—and that is all they are interested in. Sometimes, perhaps most of the time, a meme will not only be great at making copies of itself, but it may also increase the fitness of the individuals spreading it. When that happens, memes should spread even faster, but the critical point is that it *need not* happen.

A meme that is adept at making copies of itself may sometimes decrease the classic genetic fitness of the individual(s) spreading the meme (consider the yam example in the previous chapter). The celibacy meme of many religions is often touted as the best example of this phenomenon, but it is only one of many cases that have been made for memes that harm biological fitness. As Dawkins notes, "A meme has its own opportunity for replication, and its own phenotypic effects and there is no reason why success in a meme should have any connection whatever with genetic success."

It can't be overemphasized how critical it is to cultural evolution that one recognizes that memes are replicators. Dawkins notes over and over that many of his colleagues are willing to sign on to the meme as the unit of cultural transmission, but unwilling to buy the idea that memes can work in opposition to genes and still stick around for long. However, once you accept that a meme is a replicator, you either abandon the argument that memes can never work in opposition to genes, or abandon the utility of the replicator concept—which would be a huge mistake. It is that simple. What makes memes so special is that they are replicators. And if they are replicators, they increase or decrease in relation to their success at making copies of themselves, not copies of genes.

Meme Again

THE GHOST OF ADAPTATIONS PAST

When choosing among mates, human males are often very concerned with how recently a woman became sexually mature, which is a sign of fertility, and women are more concerned with the resources to which a male has access. Of course, it is not politically correct to say these sorts of things, but David Buss has found that both of these phenomena hold true in the thirty-seven human cultures that he has surveyed around the world. Such behavior is easily accommodated by standard natural selection thinking and apparently needs no further explanation. Many aspects of human mate choice, however, are not so easily explained.

While it is certainly true that we choose our mates partly on traits critical to biological fitness, many aspects of the "perfect mate" appear to make no sense in terms of genetic adaptations. We seem, at times, to be attracted to traits that are, in fact, maladaptive in terms of genetic fitness. Take a look at your spouse across the room. You certainly love her, and she is no doubt beautiful in your eyes. But in all honesty, she is also probably not the best you that could have done, if your sole criterion was genetic fitness. So how do we explain this?

There are at least two approaches to take: that of memes and that of evolutionary psychology. The meme approach is simple. If a particular meme is good at spreading itself, it will become more and more common in a population. So, for example, if the "marry overweight men" meme is adept at making copies of itself, then it will spread. "Marry overweight men because they are nice, and will support you," it might be argued, is a more likely meme. It makes no difference if the meme is accurate—it doesn't matter if fat men are nice and generous—only that the meme makes copies of itself. It doesn't even matter if marrying an overweight man means you end up with fewer kids. The "marry overweight men because . . ." meme is not concerned with honesty or genetic effects, only with making copies of itself.

Evolutionary psychology explains behaviors that appear to be maladaptive as the product of a human psyche that evolved primarily during the period when most humans lived in hunter-gatherer societies, primarily in the Pleistocene era (from 2 million years ago to 100,000 years ago). Perhaps the foremost champions of this view are Jerome Barkow, Leda Cosmides, and John Tooby. They argue:

> The most reasonable default assumption is that the interesting, complex functional design of the human mind evolved in response to a hunting and gathering way of life. Specifically, this means that in relating the design of mechanisms of the mind to the tasks and demands posed by the world, "the world" means the Pleistocene world of hunter-gatherers. . . . We cannot rely on intuitions honed by our everyday experiences in the modern world. Finally, it is important to recog-

> nize that behavior generated by mechanisms that are adaptations to an ancient way of life will not necessarily be adaptive in the modern world.

Our brains are set up to handle Pleistocene hunter-gatherer-like conditions, the argument goes, so how surprised should we be that today, in a world completely divorced from hunter-gatherer selection pressures, we often do things that provide no selective advantage?

These two approaches to understanding behavior that is currently genetically maladaptive are strikingly different. When a behavior is maladaptive from the perspective of classic natural selection, memeticists argue that this is the result of memes' doing their thing, while evolutionary psychologists claim such behaviors are the response of a well-designed mind—just a mind that happens to live in a world for which it was not designed. To understand fully the evolutionary psychology argument, however, we need to step back and examine two concepts: domains of learning and Darwinian algorithms.

Psychologists have long argued over whether the brain is "domain specific" or "domain general." This debate is far too detailed to review here completely, but it can be summarized rather easily, particularly with the liberal use of computer metaphors. The notion that the brain is domain general suggests that the brain is an all-purpose learning machine, a sort of jack-of-all-trades, when it comes to learning. The brain, they argue, generally handles very different problems using the same basic protocol. Learning in the context of foraging, sex, aggression, or cooperation, for example, is handled in the same fashion. What is learned may be different, but the brain doesn't have separate domains that tackle different learning problems in different ways. Turning to computer metaphors (a popular thing to do these days when speaking of the brain), it is as if the human brain contains one, albeit complicated, algorithm that is set in motion when something must be accomplished.

Evolutionary psychologists are of the domain-specific school of brain metaphors. They argue that natural selection has acted to produce a brain that has many different domains, each specific to a task. When faced with a particular problem, the brain allocates the problem to its appropriate domain. Here, the brain is like a giant algorithm with a large number of subalgorithms. Since these subalgorithms are thought by evolutionary psychologists to be shaped by natural selection, they are referred to as Darwinian algorithms.

The power of the evolutionary psychology approach is that it recognizes that natural selection can assign different weights to algorithms that handle particularly important fitness matters. So, for example, foraging, sex, and aggression domains might have the most well-developed Darwinian algorithms. In addition, Cosmides and Tooby, who pioneered evolutionary psychology, argue that there are finer-scale distinctions to be made with respect to Darwinian algorithms. Not only can we picture subalgorithms for specific behaviors, but we also need to think about whether the behavior is done in the context of a "social contract" or in its absence.

Social contracts involve social interactions in which one form or another of scorekeeping is invoked. A simple social contract that my parents used to enforce centered on bar mitzvah presents. How much of a bar mitzvah gift we gave someone was directly related to how much of a gift they had shelled out when they attended my bar mitzvah. My parents kept score (not exactly, mind you, but close), and the scorekeeping was in a social context. I can certainly imagine that my parents have had to keep score of lots of things in life and that many of these things are not social scorecards. The Darwinian algorithm approach argues that for scorekeeping problems of equal difficulty, people should do better on those that focus on social interactions, for this is the algorithm that natural selection has most finely tuned.

Evolutionary psychologists argue that domain specificity was determined in our hunter-gatherer past. When we act in maladaptive ways today, it is simply a consequence of our brains' having finely tuned programs; it's just that they were finely tuned to handle an environment that no longer exists. If, however, one thinks of memes as significant evolutionary forces, then whether an action is adaptive in terms of fitness is irrelevant if there is a meme hiding under the covers. If a behavior is meme driven, then we should be focusing not on genetic fitness, but rather on mimetic fitness (how many copies of itself a meme can make). Clearly these are very different views of the world. Alas, however, it is probably too early to say with absolute certainty whether the evolutionary psychology approach or the meme view of culture better explains what seems to be maladaptive behavior in humans or, for that matter, other animals.

BLACKBIRD AND GUPPY MEMES?

Memes don't have the mathematical backing that many of the models in Chapter 4 had. They have brought the notion of a kind of cultural evolution that is not completely reliant on genes to the forefront, and for that alone we should be grateful. The next question about memes is whether only folks interested in humans should take them seriously, or whether memes have been part of the cultural landscape in other species as well. The most vocal spokespeople for memes now argue that they are almost exclusively a human phenomenon (although they mention a possible animal example here and there). To bring this argument to the forefront, we must visit the work of today's most eloquent meme spokeswoman, Susan Blackmore.

Blackmore's *The Meme Machine* is the most ambitious treatise to date on memes. This book puts forth the intriguing, and controversial, claims that human language, why people talk so much, how we choose our mates, why we are altruistic, and why we have big brains are all best explained by memes (and their interactions with genes). Furthermore, religion, the art of writing, the adoption of books as a means for conveying information, and the Internet are, Blackmore argues, just ways that memes make more copies of themselves. Finally, in a logical, stunning last chapter, Blackmore claims that individuality and free will are myths: "Our Memes is who we are."

I don't intend to work systematically through all the things that memes are supposed to be responsible for in humans. Rather, I wish to examine *The Meme Machine* in terms of what it has to say about the origin of memes in nonhumans. Whatever opacities exist elsewhere in her book, Blackmore is as clear as she can be on the question of memes in nonhumans. Keeping in mind that imitation is critical to the concept of memes, consider the following:

> The thesis of this book is that what makes us different . . . is our ability to imitate.

> If we define memes as transmitted by imitation then we must conclude that only humans are capable of extensive memetic transmission.

> Memetic evolution means that people are different. Their ability to imitate creates a second replicator that acts in its own best interests.

These claims shouldn't be taken lightly, as they forced Dawkins to reconsider his views on animals and memes. In discussing Blackmore's work, he muses, "Could imitation have been the key to what set our ancestors apart from all other animals? I would never have thought so, but Susan Blackmore in this book makes a tantalizingly strong case."

Why the strong sense that animals lack memes? Blackmore's claim seems to center on the term *imitation*. For Dawkins and Blackmore, in order for memes to work, they require the organisms they inhabit to imitate. No ability to imitate, no memes. What then is the problem with memes in nonhumans? After all, I have been claiming that animals imitate and copy each other's mate choice. So, do they imitate or not? I hate to say it, but it depends on what you mean.

There is no doubt that many species of animals are capable of social learning—that is, learning from others. But a small group of influential (and very vocal) psychologists have decided that social learning is too broad a concept and that they need to subdivide social learning into many categories. These categories include "copying," "contagion," "social facilitation," "local enhancement," "stimulus enhancement," "goal emulation," "socially mediated aversive conditioning," "observational conditioning," "matched dependent behavior," and "true imitation." I will spare you the agony of working through all these.

And even after this subdivision process, there are still arguments about what constitutes "true" imitation. Since, however, we are examining Blackmore's claim that there are virtually no animal memes, it seems only fair to use her definition of imitation (or, more precisely, the definition she has adopted from others) to question her claim about animal memes. For Blackmore, "Imitation necessarily involves: (a) decisions what to imitate, or what counts as 'the same' or 'similar', (b) complex transformations from one point of view to another, and (c) the production of matched bodily action."

In fact, under this very strict definition, there may be no rock-solid cases of animal culture. But there are certainly many cases of other types of social learning, and this is all that we really need for memes to be present in animals. Before delving into animal memes, it is important to recognize that the idea that memes are a human phenomenon centers on Blackmore's adopting the strict definition of imitation. If one wishes to employ the definition that Blackmore appears to adopt, then all claims to animal imitation are questionable, and so too is the notion of memes in animals. It is very difficult to demonstrate that the three criteria Blackmore lists are satisfied—even in humans! The trouble is that while Blackmore appears to embrace this strict definition at some points, she is less clear about what constitutes imitation at other times. For example, she notes early on in her book:

> I will also use the term "imitation" in the broad sense. So, if for example, a friend tells you a story and you remember the gist and pass it on to someone else, then that counts as imitation. You have not precisely imi-

> tated your friend's every action and word, but something (the gist of the story) has been copied from her to you and then on to someone else. This is the "broad sense" in which we must understand the term imitation. If in doubt, remember that something must have been copied.

I like that definition of imitation and think it makes a lot of sense (especially regarding copying); it just doesn't happen to meet the strict imitation criteria that Blackmore uses elsewhere. Indeed, references to "broad sense imitation" (and similar phrases) can be found in numerous places throughout *The Meme Machine.* Since her thesis that memes are a rarity in animals is based on the strict definition, Blackmore's argument with regard to animals is not well founded.

We can now move on to perhaps a more critical question: Does it really matter if animals meet the strict criteria for so-called true imitation anyway? That is, can memes exist in animals in the absence of strict imitation? Are other forms of social learning enough to support meme acquisition in animals? I believe that the answer is yes.

Let's return to the basics of what makes memes replicators as Blackmore defines them. Replicators possess fidelity, fecundity, and longevity. Do animal examples of social learning that fail to meet the strict definition of imitation nonetheless show these three elements? If the answer is yes, then memes can be found in nonhumans, as well as in humans.

Consider the case that Eberhard Curio and his colleagues examined: antipredator behavior in European blackbirds. Note that the key to this story is that friarbirds are not predators of blackbirds, and normally no blackbird believes that they are when they encounter them.

Curio staged an experiment in which blackbirds could observe other blackbirds and their response to predators. In a nifty bit of chicanery, Curio set the experiment up so that an observer bird saw another blackbird display antipredator behavior in the presence of a friarbird. What had actually occurred is that the model bird was responding to a true predator (an owl), but a series of partitions made the observer *believe* that the model was responding to a frairbird. So, what does our observer do? When placed near a friarbird, it responds with antipredator behavior. Curio and his colleagues also demonstrated that this "friarbirds are predators" concept can spread through a group of at least six other birds.

There is no doubt that Curio's experiments do not qualify as true imitation, as Blackmore occasionally defines it. But they clearly are an instance of copying behavior, and I'd argue that "friarbirds are predators" should qualify as a meme. It meets the three criteria for a meme or replicator. Curio demonstrated that the information in "friarbirds are predators" is accurately transmitted through a population, so we have fidelity. Copies of this meme spread, and they'd probably have spread more if the experimental protocol was slightly different, so "friarbirds are predators" is fecund. It is hard, of course, to say anything about longevity based on a lab experiment, but there is no reason in principle that this meme wouldn't stick around for a long time if it were established in natural populations. This is especially true if this meme first took hold in older, dominant individuals. If we have fidelity, fecundity, and longevity, we have a replicator by Blackmore's definition.

What about memes in animal mate choice? To address this, let us return to my long-time lab mate, the guppy. Suppose I rig an experiment in which I have lots of observer female guppies see other females choosing very yellow male guppies. Let's further suppose our females come from a population in which there are normally few, if any, yellow males. Now suppose I take all my observer females and introduce them to a pool with other females from their own population, plus a variety of different colored males. When I then come by later, I see that not only do my original observers prefer yellow males, but so do many of the other females.

Despite the fact that females don't meet our criteria for strict imitation, we can ask the question of interest: Do we have a "yellow is sexy" meme here? I'd argue that the answer is probably yes. "Yellow is sexy" makes copies of itself, and it does so with some fidelity (I don't see many females choose, say, blue males). Again, longevity is hard to measure in controlled environments, but I could easily imagine yellow males increasing in frequency along with the "yellow is sexy" meme. Sure, it is true that *if* choosing yellow males decreased biological fitness, it's *possible* that our "yellow is sexy" meme might disappear in the end, but that is true for memes in humans is well. All that matters from the meme's perspective is making more copies of itself. This usually means working hand in hand with genes, but it doesn't always work that way. And in a race between what is good for the gene and what is good for the meme, memes are orders of magnitude faster than their biological analogues, and that bodes well for their ultimate success in such a race.

If any of the examples of cultural behavior in animals are found to be representative of behavior driven by memes, then we must stop thinking of memes as somehow new and somehow shackled to the human brain. It would be a radical change in the direction science has been following. In the closing pages of *The Selfish Gene,* Dawkins speculates on the potential power of memes:

> I think that a new kind of replicator has recently emerged on this planet. It is staring us in the face. It is still in its infancy, still drifting clumsily about in its primeval soup, but already it's achieving evolutionary change at a rate which leaves the old gene panting far behind.

Memes may have the power that this quotation implies, but even so, the more profound point is that they may be older, and more fundamental, than Dawkins, Blackmore, or anyone else has tended to think.

THE SONG OF INNOCENCE AND EXPERIENCE

We are ready to start putting the pieces of the puzzle together. Our innocent genes and our experience of culture each affect behavior, particularly mating behavior, and the theory underlying both genetic and cultural transmission is not all that opaque. We can now look at how these forces *interact* in shaping behavior.

6

Are You My Type?

> Hence we must all believe that all the sciences are interconnected, that it is much easier to study them all together than to isolate one from all the others. If, therefore, anyone wishes to search out the truth of things in serious earnest, he ought not to select one special science, for all the sciences are cojoined with each other and interdependent.
>
> RENÉ DESCARTES, *Rules for the Direction of the Mind* (1629)

ONE HARSH BUT CRITICAL LESSON THAT research scientists learn early on is that the scientific process can be frustrating. It may be the best process ever conceived for objectively testing hypotheses, but it is frustrating. It is human nature to attempt to solve problems in the fewest steps and the quickest fashion. Unfortunately, most of the time, such solutions cannot possibly be scientific.

When my son, Aaron, and I try to build grand towers

from simple wooden blocks, I need to remind him (and myself, for that matter) that to reach our goal, we need to remember what we hope the end product will look like, and to keep in mind what function various blocks play in producing such a product. We don't necessarily need to know exactly what every block does or exactly how every block interacts with every other block. For that matter, we don't even need to know what the blocks are made of. What we do need to understand is something about what functions various blocks serve. Then, and only then, can we construct our grand tower. Constructing theories and experiments to understand complex questions in science is not all that different.

Just as with Aaron's towers, what we can say about how genetic and cultural transmission *interact* to shape mate choice begins with an understanding of building blocks—in this case, genes and culture. Now that we have examined each of these forces separately, we can move on to their interactive capabilities. This is not to say that there isn't a great deal more to be learned about each process in isolation of the other, but rather that we arguably know enough at this point to start examining their combined force.

THE COMPLETE GUPPY

As it happens, the most readily available way to look into how genetics and culture interact in shaping mate choice is through the guppy. And as we've seen, female guppies are strongly influenced by the choices that other females make regarding sex with males. On the other hand, we know more about the genetics of guppy mate choice than most people could imagine, and innate predispositions also play a major role in determining whom a female guppy chooses as her mate. Experimentally examining the relative role that cultural and genetic transmission play in guppy mate choice is an obvious opportunity.

In examining the interaction of genes and culture in guppies in my own lab, there are a number of experimental tactics I might have adopted. For example, I could have tested what occurred when genetic and cultural transmission operated in the same direction. That is, I could have set up an experiment in which females had the chance to imitate the choice of others, when such others were expressing an innate preference to begin with. But any male both with traits that females innately prefer and with other females around him is just too perfectly attractive, and not that interesting. Take two very strong forces and combine them in a positive fashion, and you will, in all likelihood, get an even stronger force operating in the same direction—but experimentally oppose such forces, and you might find something interesting. What if information obtained by cultural transmission suggested that females should choose males with traits that were in direct opposition to their innate tendencies? In guppies, females from the Paria River in Trinidad have a genetic predisposition to mate with males that contain lots of orange body color. What if culturally derived information suggested that other females were preferring drab versus orange males?

To examine the relative importance of genetic versus social cues in shaping female mate choice, I ran what might best be called a "titration" experiment. Guppies used in this experiment were from the Paria River and hence showed a heritable preference for preferring brightly colored males, particularly brightly orange-colored males. Part one of the titration experiment involved females that observed two males. Pairs of males were either matched for the amount of orange color they had on their bodies, or pairs of males differed by 12, 25, or 40 percent in amount of orange color. Armed only with this information, females chose randomly between males matched for orange and virtually always chose the more orange of a pair of males.

Part two of this experiment involved a bit of manipulation on my part. Using a series of see-through Plexiglas partitions, an observer female always observed another female choosing the less orange male (of a pair) as her putative mate. Again, males were either color matched or differed by 12, 25, or 40 percent total body color. Recall that with no model present, flashy orange males were the clear winners in such tests. But what did our observers do after witnessing another female choose the drab male of a set? Before answering this, consider the evolutionary drama unfolding here. A female's genetic predisposition is pulling her toward the more orange male, but social cues and the potential to copy the choice of others are tugging her in the exact opposite direction—toward the drabber of the males.

What happened in this evolutionary tug-of-war depended on a single variable: just how different males were in orange coloration. When males differed by small (12 percent) or moderate (25 percent) amounts of orange, females consistently chose the less orange of the two. Here, the tendency to copy overrode a genetic predisposition for orange males. If, however, males differed by large amounts (40 percent) of orange, females ignored the choice of others and preferred oranger males. Here, genetic predisposition masked copying effects. Loosely speaking, it is as if a threshold existed somewhere between males differing in orange by 25 to 40 percent, and below this threshold social cues were predominant. Above this threshold, however, genetic factors could not be overridden.

It was the first experiment ever done that revealed interaction of genetic and cultural transmission in shaping female mate choice. The results were somewhat surprising, in that rather than one force always predominating, a threshold effect was uncovered. The most important aspect of the study was the general applicability of the protocol. Titration is a promising tool for examining the relative importance of social and nonsocial cues in many contexts and across many species. For example, although the focus was on the role of genetic and cultural transmission in shaping mate choice, the idea of opposing these forces with respect to any behavior is now plausible. If aspects of foraging behavior or aggression or whatever other behavior one is interested in have innate components, but are also influenced by cultural variables, the protocol developed in the titration experiment can be used to decipher complicated behavioral scenarios.

One nagging question after completing the titration experiment was whether there was anything that could be done to make the very drab males (40 percent less colorful) more desirable. I ran an experiment similar to that just described. Once again, I had females choose between two males that differed by 40 percent orange coloration. As expected, when no model was around, females preferred the flashy orange male. But now, instead of showing a female a single model near the drab males, I instead showed her two models in a row preferring the drab male. Sure enough, that made all the difference, and drab males were suddenly preferable to flashy ones. These results speak volumes about the power of cultural transmission even in systems where genetics plays a key role. It further suggests that the threshold observed in the prior experiment was itself flexible.

An important question, however, lurked below the surface and was often raised during seminars I gave on this topic. All the work compiled had clearly shown that females copied the mate choice of others, but not that such copying effects truly changed their long-term sense of what *type* of male was attractive and what *type* wasn't. For example, it was clear that if male 1 was near a model female, male 1 would become more attractive in the eyes of an observer. But a stricter view of what constitutes cultural transmission might require that not only male 1, but other males possessing traits like that of male 1, would also be perceived as more attractive by an observer.

Within this view of cultural transmission, males with the same *phenotype* as male 1 all become more attractive to an observer. If male 1 is blue-colored, then all blue-colored males, not just male 1, become more attractive in the eyes of our observer female. Here, an observer's "gestalt" view on what makes a male attractive has been changed by her opportunity to copy.

Together with Jean-Guy Godin and Emily Herdman, I set out to see if our guppies and their mate copying fit into this stronger, gestalt view of cultural transmission. The protocol, as in so many of the other mate-copying experiments, required many controls, but the basic experiment was not very complicated. We reasoned that we could set up an experiment with drab and bright Paria males and determine if mate copying could make *all* the drab males—not just the ones near the models—more attractive. If we put a model near a drab male and that made him more attractive, we could then easily test not only whether an observer female liked the drab male she observed, but whether in the eyes of such a female, drabness *itself* was now a desirable trait for males to possess.

Not surprisingly, what we found was that it was easy to place a model female near a drab male and make that male more attractive. We then took our observer female and placed her between a new set of drab and colorful males and examined which of these males she preferred. Our result—the observer now found drab males in general more attractive—was clear evidence that mate copying constituted cultural transmission not just about an individual, but about types of individuals.

THE BIZARRE LOVE TRIANGLE OF THE SAILFIN MOLLY

Like the guppy, the sailfin molly is a tiny fish with a big story to tell: a story about the interaction of genes and culture in shaping mate choice. The story begins with a bit of basic biology. When given the option, sailfin molly females, like females in most other species of fish, prefer mating with large males. Cathy Marler and Mike Ryan found that when males differed in size, female sailfins were quite consistent in their choice of the larger of two males. Why they have such a preference is still open for debate, but Marler and Ryan argue that it probably revolves around a nervous system that is generally constructed to hone in on large objects (large food items, large predators, large potential mates). That is, larger males are not better than smaller ones; they're just more obvious. Furthermore, whatever the reason underlying this preference turns out to be, Marler and Ryan provide indirect evidence that the preference was passed down from an ancestral species (a species that probably lived at least 10,000 years ago) and hence probably has a genetic basis.

Using the basic protocol employed in the guppy experiments, Katherine Witte and Mike Ryan examined whether females copied each other, given that a pair of males was equal in size. By matching for size, they removed the effect of innate preferences—no differences in male size, no reason for females to prefer one male over the other based on innate predispositions. Once a female then saw another female near one of the two same-sized males, it was clear that female sailfins copied each other's choice of mates. Witte and Ryan then pitted innate preferences against cultural transmission by placing a model near the smaller of two males. Just as in the guppy titration experiment, culturally obtained information "pushed" the fish in one direction (toward the small male), while innate preferences "moved" her in the opposite direction (toward the larger). Results were straightforward in that female sailfins still preferred the larger male; so although cultural transmission is important in this system, it can't mask any innate preferences the fish may already possess. Yet the story of sailfin mollies, genes, and culture is hardly over. Rather, we are about to introduce a new, very odd actor onto the stage: the Amazon molly fish.

You may think you have heard of every strange love triangle there is, but you haven't heard anything until you've heard about the Amazon molly (*Poecilia formosa*) and its friends. This is a species of all female fish. No males—not one. Individuals in this species reproduce by creating clones of themselves. That is, females produce eggs that are not fertilized by sperm, but still develop into mature new females for the next generation. This sort of reproduction is more common in animals than most people think.

What makes Amazon mollies particularly interesting is that not only do they have no males, but they are "gynogenetic." What this means is that although their eggs are never fertilized by sperm, females must still mate and obtain sperm from a male from another species! The unlucky males are sailfin mollies. The sperm from sailfin males never unite with the eggs from Amazon mollies, but they somehow or another stimulate egg development. Once the sailfin sperm initiate egg development, they are essentially tossed out by the Amazon females, never to make it to the next generation.

The sailfin sperm spilled in mating with an Amazon molly are wasted. They never fertilize an egg, and so are an evolutionary dead-end. Why didn't natural selection put the brakes on male sailfins' mating with Amazon mollies long ago? After all, wasting copies of genes is not the sort of thing natural selection usually favors. In an ingenious experiment, Ingo Schlupp, Cathy Marler, and Michael Ryan found that the answer to this paradox lies in mate copying.

It turns out that male sailfin mollies are not completely wasting their time when they mate (but never fertilize) female Amazon mollies. The reason this effort is not for naught is that female sailfin mollies are often watching off in the distance. As we just learned, female sailfin mollies find males that mate with others more attractive and go for such males themselves. This is true even when males are mating with females from *another species.* Female mate copying on the part of sailfin mollies, then, not only affects mate choice in that species, but appears to be the glue that holds together this strange three-way love triangle.

WHALE CULTURE

I'm not sure exactly why, but people expect great things from whales. Of course, they are the largest mammals in the world, and so they have the biggest brains, but what that means about how and what they think is unclear. The issue of the relationship between body size and intelligence remains a contentious one, and if the facts we have uncovered demonstrate one thing, it is that very simple animals can behave in very complex ways. Big-brained need not mean intelligent. Then there is the Moby Dick complex that whales are smart because each one seems to have a unique personality. No doubt this is true, but I'm not sure that whales have unique personalities any more than my charming guppies do.

Despite the fact that much of the lore surrounding whales is based on anecdote and anthropomorphism (projecting human emotions to nonhumans), there is now good evidence that culture plays a large role in whale life—a role that interacts with whale genetics in a new and fascinating manner.

When he is not doing mathematical models (and maybe even when he is), Hal Whitehead is out on boats studying pilot whales, sperm whales, and killer whales. Two commonalities found across these species are that females spend their lives with other females that are relatives and that new groups tend to form via fissioning of an already existing group. Such a social structure is technically referred to as matrilineal. In addition to the behavioral characteristics that define matrilineal societies, there is a genetic correlation. It turns out that matrilineal whale species have much less genetic diversity than other species of whales.

Using data accumulated by others, Whitehead found that the DNA in matrilineal species is ten times less diverse than in other whale species. What is particularly interesting is that this is true only for a specific kind of DNA called mitochondrial DNA (mtDNA). While normal DNA is inherited from both mothers and fathers, mtDNA is inherited only through mothers. Naturally, Whitehead wanted to know whether there was any link between the fact that mtDNA was inherited through mothers and that it was least diverse in matrilineal species. What he found was an unexpected but intriguing interaction between genes and cultural transmission.

To get an appreciation for how genes and culture interact in whales, we need to step back and review a basic concept in population genetics: the notion of "hitchhiking" genes. The easiest way to envision hitchhiking is to imagine two genes, let's call them 1 and 2, with each gene having two variants, so that we have four total varieties: 1a, 1b, 2a, and 2b. Now, suppose that these two genes are linked in the sense that when you find 1a, you find 2a (and similarly for 1b and 2b). We then have a situation in which the fates of genes 1 and 2 are not independent; what happens to gene 1 can have implications for gene 2, and vice versa.

Continuing with our discussion of hitchhiking, consider the case in which variety 2a is much superior to 2b. Natural selection will clearly cause an increase in variety 2a. But what if varieties 1a and 1b are neutral, in that natural selection does not favor either type? Normally, this would translate into chance determining which variety (1a or 1b) we observe. If they're equally good at solving some problem, then sometimes by chance we will see 1a and sometimes 1b. Since hitchhiking is occurring here, however, we would expect something quite different. Although 1a and 1b are neutral, we still expect to see 1a most of the time because it is associated with 2a, and 2a is increasing in frequency due to selection. Thus, 1a increases in frequency as a result of hitchhiking a ride with 2a.

Returning to whales, Whitehead invokes a new type of hitchhiking to understand the drop in genetic diversity seen in matrilineal species. Instead of postulating that two genes are linked, Whitehead suggests that mtDNA varieties are linked to cultural traits in matrilineal species of whales. We know that cultural transmission plays a significant role in whale society. Young females, who will spend most of their lives with older female relatives, learn a lot from such relatives. For example, cultural transmission appears to play a significant role in foraging behavior, as well as in the distinct "dialect" used in vocal communication in whales. It is not hard to imagine, based on song learning in birds, that dialect has implications for mate choice (our link to mating). Now suppose that one variety of a culturally acquired trait in whales was superior to another variety. Then we would expect the superior variety, even if culturally inherited, to increase in frequency. If we started out with many cultural variants, we would end up with one—a decrease in diversity.

Whitehead postulates that somehow mtDNA variants are linked to culture variants. If the mtDNA variants are neutral, as assumed in Whitehead's model, then as diversity in cultural variants decreases, so too would mtDNA diversity. As a result of hitchhiking, decreases in cultural variation lead to decreases in mtDNA diversity, and we have our gene-culture interaction.

It is worth noting that Whitehead's findings are radical. In this age of molecular biology, many research labs study mtDNA, a case in point being that mtDNA was central to the "Eve" hypothesis of human evolution. The Eve hypothesis suggested that humans share a single common female progenitor who lived about 200,000 years ago. It was mtDNA work that underlay all the work on Eve, as well as many other studies ranging from work on human evolution to medical issues. Before Whitehead's work, the notion that mtDNA evolution may be linked to cultural evolution would have been dismissed out of hand.

SINGING OUT TO SAVE YOUR GENES

The stomping grounds of Peter Grant and Rosemary Grant are the Galapagos Islands, and their intellectual godfather is Charles Darwin. In addition to their adventures (associated with twenty years of living part-time on a set of small, drab brown islands that also happens to be a tourist trap), the Grants, in collaboration with dozens of colleagues, have collected data on virtually every aspect of the life of the myriad of finch species inhabiting the Galapagos. Their long-term study is already a landmark in behavioral ecology and demonstrates the power of evolutionary thinking when examining everything from physiological to behavioral questions.

Among the many problems the Grants have tackled is the role of cultural transmission in the evolution of finch songs. As we saw in an earlier chapter, cultural transmission is part and parcel of birdsong evolution, but as usual, the Grants have found a new twist in the way this operates in the finches of the Galapagos. In the Galapagos finches, cultural transmission not only underlies birdsong, but interacts in a new way with the genetics of reproductive isolation in these birds.

The medium ground finch (*Geospiza fortis*) and the cactus finch (*G. scadens*) both inhabit the Galapagos island of Daphne Major. All species of finch on the Galapagos probably diverged from a common ancestor about three million years ago and, not surprisingly, look quite different from one another. What is surprising is that although they are considered to be different species, numerous cases of interbreeding between finch species are well documented. Yet despite the interbreeding, finch species manage to maintain their integrity. How is this possible?

Without barriers blocking interbreeding, why don't the many species of finches just merge into one species? This is particularly vexing because if there are no defined barriers blocking interbreeding, one often finds that the offspring that arise from such a gene swap are inferior in quality. Not so for the ground finch and the cactus finch, and yet somehow they remain two species.

The Grants found that cultural transmission of song in both the ground and cactus finch may help explain why even though interbreeding is possible and apparently not costly, it isn't common. What they discovered was that the songs transmitted across generations in these two species were quite different from one another. For example, the song of the cactus finch has shorter components that are repeated more often than that of the ground finch. These differences have a dramatic impact on gene flow across species. Of 482 females sampled, the vast majority (over 95 percent) mated with males who sang the song appropriate to their species. Cultural transmission of song allows females to recognize individuals of their own species and provides a barrier to forming hybrid offspring.

The proof of the pudding often lies in the exception to

the rule, and this holds true for the Grants' argument regarding song and gene flow. In their twenty years of study on Daphne, they found eleven instances in which the male of one species sang the song of another species. In one instance, a young male had an overly aggressive male neighbor from a different finch species and took on the song characteristics of that neighbor. The result in such instances is cross-species breeding, resulting in a hybrid. Remove the normal pattern of cultural transmission, and the barrier to breeding across species disappears.

The Grants' foray into how cultural and genetic transmission operate doesn't end with their work on gene flow and birdsong. Far from it. Instead, they tackled another important genetic issue in evolutionary biology—inbreeding—and once again showed how cultural transmission interacts with genetic transmission in new, unexpected ways. In most cases, breeding with relatives (inbreeding) tends to be selected against, the basic reason being that it increases the probability that deadly illnesses that require an individual to have two copies of a "recessive" gene are much more likely to result from such matings. It turns out that when the Grants examined how ground finches avoid the negative genetic consequences of inbreeding, the answer lay in cultural transmission of song.

There are many cues that animals could potentially use to recognize relatives and avoid mating with such individuals. In ground finches, the Grants discovered that males sing a song very similar to that of their father and their paternal grandfather (their father's father). A male's song tells something of his heritage. Females use this genealogical information on males to determine who to mate with and, at the same time, how to avoid inbreeding. Ground finch females simply won't mate with males that sound like their own father. So, once again, cultural transmission via song learning interacts with genetic transmission in a rather unique and interesting fashion.

There is something intuitively appealing in the fact that such a nice example of the interaction of genes and culture comes from the set of islands made famous by evolutionary icon Charles Darwin. The Grants' work on finches' songs and genes has certainly done Darwin justice.

A NUDGE OVER THE TOP

The concept of "fluctuating asymmetry" is the rage in both evolutionary and psychological work on mate choice. The work on fluctuating asymmetry is the result of a long search for general traits that can be used as indicators of the overall health and fitness of an individual. Wouldn't it be wonderful, the argument goes, if there was a simple cue that individuals could hone in on and use to determine the underlying genetic quality (with respect to health, resistance to disease, and so forth) of potential mates? Fluctuating asymmetry may be such a cue.

Evolutionary biologists focused their search for some

measure of health and vigor by examining what is called "developmental stability"—how well an organism handles changing environments as it matures. The underlying premise here is that organisms that can handle changing environments well may be those favored by the process of natural selection. Suppose that individuals do in fact differ in their proclivity to respond to the changing (and often adverse) conditions they face throughout their development. How would anyone else be able to tell who is good and who isn't? Bodily symmetry may be one means.

Symmetry is a measure of the similarity of the left and right sides of an organism. For example, if your right and left arms were exactly the same length, your arms would be perfectly symmetric. One way of measuring this would be to subtract the length of the left arm from that of the right. In that case, perfect symmetry would produce a score of zero, and any score greater or less than that would indicate some asymmetry. If others could assess your degree of asymmetry for a particular trait, they could presumably use that information in their mating decisions, and if symmetry is a good measure of developmental stability, using it as a cue should be favored by natural selection.

The "fluctuating" in fluctuating asymmetry refers to the fact that any deviation from a score of zero is bad. In our example, it doesn't matter if you have too long a left arm or too long a right arm, just that your arms are not the same length. The expectation is that the amount one sees of low or high scores will *fluctuate* over evolutionary time.

Differences in symmetry across individuals represent, in some indirect manner, differences in their overall quality. The argument goes that individuals who choose mates that are symmetric with respect to some trait are in reality choosing mates with a high overall genetic quality. In this light, in an extensive review of fluctuating asymmetry in forty-two species ranging from fruit flies through ducks to humans, Anders Møller and Randy Thornhill note, "It has been predicted that, in general, across species, there will exist a biologically significant negative correlation between asymmetry and success of individuals in sexual competition." In order to test whether this very general and potentially quite important statement holds water, Møller and Thornhill used meta-analysis, a statistical technique.

Rather than running an experiment or completing a typical review paper on a subject, meta-analysis entails taking the results generated across a large number of studies and using newly developed statistical tests to examine whether there are any underlying trends across *all* such studies. Of course, the hard work and detailed biology of each study are glossed over in a meta-analysis, but the resulting ability to uncover large-scale trends more than makes up for this, or so argue the proponents of this approach.

Møller and Thornhill's meta-analysis used the results of 162 studies of fluctuating asymmetry. These studies ran the gamut in terms of what traits were examined and how they were examined. Some were done in the lab, some in the field. Some looked at asymmetry in specific traits (such as facial asymmetry), while others measured asymmetry across a wide variety of traits. Given this hodgepodge of data, Thornhill and Møller searched for some underlying generalities regarding symmetry. And they found three such underlying truths.

With respect to their main prediction—more symmetry translates into greater mating success—the results were unambiguous. A clear and powerful relationship exists between degree of symmetry and attractiveness, and it is precisely in the direction predicted. In most cases, more symmetric individuals were chosen more often as mates. Second, and not surprisingly, Møller and Thornhill found that when symmetry was measured in traits that evolved in the context of mate choice, the correlation between symmetry and attractiveness was stronger than it was in the case of other traits. Finally, in humans, facial asymmetry seems to be the trait that was most strongly related to attraction.

Møller and Thornhill stress two other aspects of symmetry relevant to our purpose. First and foremost, there is evidence that the degree of symmetry displayed by a given individual has a genetic component in that it is heritable. In addition to the heritable nature of symmetry, there appear to be considerable benefits to choosing the most symmetric mates possible. For example, when choosing symmetric males, females are more resistant to parasites and tend to produce more and healthier offspring.

Given that differences in symmetry are heritable, that

symmetric individuals are more attractive, and that individuals gain much by choosing symmetric partners as mates, we can propose a few experiments that would further elucidate the interactive nature of genetic and cultural transmission in shaping mate selection. To do so, we will use symmetry as the genetic trait and examine how it might interact in two systems with well-known cultural components: mate copying in birds and in humans. First, the birds.

In our review of cultural transmission and mate choice, a strong case was made that Jacob Höglund and his colleagues have demonstrated mate copying in black grouse. By using an elaborate set-up involving stuffed female birds placed on the territory of males, Höglund's group was able to show that such males became much more attractive in the eyes of assessing females. How might this sort of cultural transmission interact with genetic factors such as fluctuating asymmetry in the black grouse? To answer that question, recall that on black grouse arenas, a single male will often obtain on the order of 80 percent of all copulations. It turns out that the position of a male's "territory" in the arena is another cue that females use when deciding which male will be the lucky male. Males that end up with territories in the center of an arena, where territories are small and males tend to cluster, are much more likely to be the ones to obtain an inordinate percentage of mating opportunities.

Given that territory placement is so critical in male mating success, it shouldn't be all that surprising that Höglund and his colleagues have been trying to figure out just what determines which males end up with the coveted central territories. A number of factors appear to be critical in such a race to the center, and sure enough, bodily symmetry appears to be underlying many of them. Males on center territories are indeed more symmetric than males on the edge of an arena, and this translates into greater reproductive success. Symmetry appears to be a very good cue for a male's genetic quality, as more symmetric males (at the center of an arena) have higher testosterone levels.

So males in the center of a lek are more symmetric, and females prefer such males (apparently for good reason). Layered on this is cultural transmission in the form of mate copying. To examine the interaction of these forces in shaping mate choice, I propose two experiments. In the first, which would be a controlled study most likely run in a big, open aviary, one begins by allowing the grouse to form their arenas. Then in one treatment, single females are placed near such an arena, and their choice of mates is recorded. I'd predict that one of the center males would receive a large proportion of the matings in this case. For argument's sake, let's say that after twenty females go through this procedure, male A gets 50 percent of all the copulations. Next, run a similar experiment, except now use pairs of females. That is, let a female make a choice, while another female observes. Then let the observer female make her choice. Now let's say that male A obtains 90 percent of all matings. The difference between the two experiments (90 percent − 50 percent) then provides some information on the relative magnitude of cultural and genetic transmission in a seminatural environment. In this case, both forces would be comparable, as cultural transmission almost doubles the mating success of a given male.

A second hypothetical experiment examining gene-culture interaction in grouse mating might look something like the guppy titration experiment. Here, we would utilize the stuffed female grouse manakins so cleverly used by Höglund and his colleagues in earlier work. But while Höglund's group randomly chose a male territory to place a manakin on, in this experiment such placement would be anything but random. The idea would be to place female grouse dummies on the territories of males that are not at the center of a mating arena. This would place females in a situation in which choosing center males would still have all the genetic benefits it did before, but cultural cues would be "pushing" a female away from such males. Just as with our first grouse experiment, one could then measure how a male on the edge of an arena fared with and without female manakins on his territory to get a sense of the extent to which cultural transmission can help a male that would be in bad shape without it.

Now on to some thoughts regarding mate copying and symmetry in humans. The literature on the strength of fluctuating asymmetry in human mate choice is as convincing as it is in animals. Although it is true that there are no direct experiments on the heritability of symmetry in humans, all indications are that it is as heritable in humans as in nonhumans. Further, there is no a priori reason to believe that a trait that is heritable when reviewing the large animal literature should fail to be so for humans. Clearly many human traits are not heritable, but these are often the ones on which we most differ from animals, not those we share in common.

In their meta-analysis demonstrating that individuals prefer more symmetric partners as mates, Møller and Thornhill included forty studies of symmetry and mate choice in humans, clearly indicating this pattern in our own species. In addition, other studies demonstrate that human mate preferences are actually quite consistent the world over, and a strong case can be made that symmetry plays a significant role in human choice. In fact, one of the more interesting studies on just how important symmetry is in the context of mate choice was conducted in humans.

In a 1998 study, Steve Gangestad and Randy Thornhill posed the following question: Do females change their reliance on symmetry as a function of menstrual cycle? Their hypothesis was straightforward: if symmetry is important to females as a way of assessing males, then they should be much more attuned to this when ovulating than at other times during the month. To test this, Gangestad and Thornhill began by giving women T-shirts worn by different males because, believe it or not, women can make a fairly good guess at male symmetry simply by smelling the odor that a male emits. Exactly why this is so is still not understood (although it probably harks back to our earlier discussion of MHC and odor), but given that it is, T-shirts are a reasonable experimental tool.

Using forty-one college women as subjects, Gangestad and Thornhill distributed "smelly" T-shirts that had been worn by males for two days. Females were asked to rank the scent of a given T-shirt in terms of the attractiveness of the donor. Gangestad and Thornhill also measured the symmetry of the male T-shirt donors and the reaction of females to their T-shirts at different periods during a normal ovulation cycle. Sure enough, women showed a stronger preference for the T-shirts of symmetric males just at the time when they were most fertile (that is, the time when conception would be most likely). Other work has shown that both men and women viewed more symmetric individuals as more attractive, dominant, sexier, and healthier.

The fact that individuals prefer symmetric mates may help explain a recent study by Jeffrey Simpson and his colleagues that found differences in how symmetric and asymmetric males behave when competing for females. Simpson and his co-investigators staged the following scenario. An attractive woman interviews two males who know they are competing for the chance to date her. The males are then told they must stand before a video camera and tell the pretty interviewer, as well as their competitor, why they are in fact the better choice for an evening on the town. It turns out that symmetric males are much more likely to take an aggressive tack and make direct comparisons between themselves and their rival than are asymmetric males.

Presumably this is because symmetric males have had more success in the dating game than their asymmetric peers, and are thus more comfortable believing they will inevitably attract a female after comparing themselves directly to others. In any event, females like symmetric males, and this factor appears to shape male sexual strategies as well.

Not only do humans prefer symmetric partners, but choosing symmetric individuals may produce great benefits in humans, just as in nonhumans. Three lines of evidence—one direct and two indirect—suggest this. The direct evidence centers on more symmetric individuals' having a reduced chance of illness and other related medical problems, including exposure to toxins and infection, schizophrenia, preterm births, and Down's syndrome. In terms of indirect benefits, symmetric individuals have higher IQ scores, and females report more orgasms with symmetric versus asymmetric sexual partners.

Given that symmetry in humans is probably under genetic control of some sort and that symmetric individuals are more attractive and perhaps more fit, how might we examine the role of this variable in conjunction with cultural transmission? To begin with, recall that people "date copy" (see Chapter 3). That is, if someone of the opposite sex is rated highly by others as a potential date, this influences the choice of those presented with that information. Of course, when experimentally examining date copying, researchers have great flexibility. For example, in my study with Michael Cunningham and Duane Lundy, we told a female subject that either zero, one, or five out of five other females found some guy—we'll call him Joe—interesting enough to date. Subjects were much more likely to consider dating Joe themselves when they believed that others would as well. What is surprising was not so much our finding as the fact that no one had done the study before.

Suppose that instead of giving people only written information about potential dates, we now also show them a series of pictures. Once pictures enter the story, we can use computer programs to manipulate photos in any manner we deem interesting. What if female subjects were shown two photos—one of a relatively symmetric face and one of a more asymmetric face? Layered on that, suppose we rig it so that females are told that a whole suite of ladies said that they were more willing to go out with the asymmetric male. What would our subjects do then? Would they choose the symmetric male, ignoring cultural information, or the asymmetric male, in opposition to almost all other studies of female mate choice? Would it matter how different faces were in terms of symmetry or how many other women a subject believed were interested in the more asymmetric male? I'd bet that if our subjects believed enough other women found the asymmetric face attractive (and it wasn't too asymmetric), they'd go that way themselves. I'd further predict that when interviewing subjects about why they chose as they did, many would say they assumed that they "missed something" about the picture that others picked out. In any case, the experiment wouldn't be all that difficult to undertake, and it might shed new light on the interaction of genetic and social factors in shaping mate choice.

BEYOND TWINS

A large body of literature has been generated by the "nature versus nurture" debate concerning the interaction of genetic and cultural transmission. The data are largely concentrated in studies of identical twins in which researchers attempt to separate genetic and "environmental" factors, including how such factors shape mate choice. For example, if identical twins are raised apart, then most of the differences between such twins is environmental, because they are genetic clones of one another. Environmental factors, however, are essentially anything nongenetic, and that is too broad a concept for effective discussion here. That combined with the fact that many of the terms in the nature versus nurture debate are ill defined convinced me to omit a discussion of such studies.

On reflection, perhaps it shouldn't be all that surprising that there are only a handful of controlled studies of how genetic and cultural transmission affect the evolution of mate choice. To begin with, the number of studies of cultural transmission (alone) and mate choice, while constantly growing, is far from huge. Only recently have such studies been recognized for their value, and they tended to be viewed skeptically at the start of the process. This isn't shocking, as new ways of framing critical ideas in a discipline are always viewed with some trepidation early on.

On the genetic side of the coin, there has certainly been a fair share of work done on mating behavior. Still, this is only a small fraction of the work done on the genetics of behavior. What is more important for our purposes is that the detailed, controlled work that has been undertaken on the genetics of mate choice tends to have been done on very small creatures: fruit flies and the like. This makes perfect sense since the smaller the critter, the quicker it tends to mate, and that means lots of generations' worth of data on the variable of interest in a relatively short period of time.

The conundrum, then, is that despite the importance of these studies, we have not all that many on cultural transmission of mate choice (or behavior in general), and the scientists who tend to get the most excited about such studies are those who work with large creatures—primates and the like. They may not do controlled experiments, but they get excited. Although we have time and again seen that cultural transmission can be important in smaller creatures (with smaller brains), those with an inherent interest in culture persist in focusing on larger animals. Moreover, the work on the genetics of behavior tends to be focused not just on small creatures, but on *really* small creatures. When you cross these two sets (the cultural and genetic studies of mate choice), you are left with a very small number of controlled studies to review.

Naturally, one needs to start the study of interaction by choosing a species whose mate choice behaviors are affected by both genetic and cultural transmission. The guppy experiments demonstrate a fairly simple, malleable procedure for what might come next in the process. After examining each force in isolation of the other, one can "titrate" one force against the other. In the guppy work, this was done by manipulating either how different the males were in color (the genetic side of the equation) or how many other females an observer saw prefer that male as a mate (the culture side of the equation). But the species need not be guppies, and the behavior of interest need not be mating to use this procedure for experimentally examining how genes and culture interact to shape behavior.

Imagine you were a researcher interested in aggression, and you knew of the guppy mate-choice work. Further suppose you study a species in which some of the genetics underlying aggression are known—for example, some primate. After studying this species for a while, you note an odd phenomenon: after individuals observe others in a fight, they behave differently paired with those they have observed. Observers, you note, seem to act in a different manner when paired up with aggressive versus submissive individuals they were able to observe. An observer is much less aggressive when paired with someone it saw winning a fight than when paired with an individual it observed losing a fight. Cultural transmission changes fighting behavior. Armed with this information, one could easily run an experiment similar to the guppy titration work. In such an experiment, one could take individuals with different innate aggression levels and expose them to a spectrum of others winning or losing fights. Run all possible combinations of genetic and cultural factors (innate aggressive/interact with winner, innate submissive/interact with loser), and deeper insight into the interaction of genes and culture would surely follow.

In principle, there are many ways one could devise to examine the interaction of genetic and cultural transmission. It should be a tremendously valuable avenue, but what we need are more active labs working on this interaction question to come up with new ways of investigating how genes and culture interact—and not just in the context of mating, but with respect to all social behaviors. At the same time, the work we have examined here shows much promise for helping us truly understand how genetics and culture interact in shaping behavior. We are learning why we are who we are at a greater rate than it seems we ever have before.

7

Animal Civilization

> Let them alone: they be blind leaders of the blind: And if the blind lead the blind, both shall fall into the ditch.
>
> MATTHEW 15:14

A FEW HUNDRED YEARS AGO, IN ORDER TO protect a diamond mine in the former Congo, some innovative Congolese figured out a mechanism for training a population of gorillas to use a new weapon to guard their treasure. The gorillas were quite good at using these weapons and continued to do so for many generations—long after the people who trained them were dead. The trouble was that the gorillas didn't want to stop using these weapons—ever—and became rather adept at wielding them. They taught their children how to use them as well. The ultimate outcome was that they posed a real threat to explorers who rediscovered the mine the gorillas had been guarding for hundreds of years. Obviously cultural transmission played a large role in this rather strange tale. After all, the folks who trained the gorillas are long gone, and clearly the gorillas have passed down their acquired behavior by teaching new generations just how to use their weapons.

Who could have predicted that animal culture could have such bizarre consequences? The answer, in case you haven't guessed it, is Michael Crichton. The gorilla case above is completely fictitious and comes straight from Crichton's novel *Congo.* From our perspective, however, Crichton's narrative is illustrative not for accuracy but because it is plausible. Given what we know about culture, Crichton's example, under the right conditions, could happen. Nature, however, is rarely outdone by fiction, and as we are about to see in our sortie into cultural transmission outside the realm of mating, many examples of animal culture rival Crichton's in their almost creepy ingenuity.

IMO THE AMAZING

Imo the Japanese macaque holds a special place in the hearts of those who study the cultural transmission of behavior. Her story starts innocently enough on Koshima Islet, Japan, in September 1953, when Imo, only a year old, added a new behavior to her repertoire: she washed the sweet potatoes that researchers provided her in a nearby brook before she ate them. Soon enough, many of Imo's peers and relatives had learned the art of potato washing from our pioneer epicurean. By 1959, most infants in Imo's group intently watched their moms, many of whom had acquired Imo's habit, and learned to wash their own sweet potatoes at early ages.

Imo was more than a shooting star that had a brief moment of glory, only to fade into obscurity. When she was four, she outdid herself by introducing an even more complicated new behavior into her group. In addition to the sweet potatoes that monkeys on Koshima Islet were given, they were also occasionally treated to wheat. The problem was that wheat was usually provisioned to the monkeys on a sandy beach, and wheat and sand mixed together is not nearly as tasty as a plain helping of wheat. So Imo came up with a novel solution: she tossed her wheat and sand mixture into the water. The sand sank, the wheat floated. Imo had done it again! As with the sweet potatoes, it was only a matter of time before her groupmates learned this handy trick from Imo. It took a bit longer for this trait to spread through the population, but, of course, it was trickier than washing sweet potatoes, at least from a monkey's perspective. Monkeys aren't used to letting go of food once they get their paws on it, so it was tough to learn to throw the wheat and sand combination into the water. But eventually this new behavioral trait was spread to many group members by imitation and cultural transmission.

Imo's escapades, and the cultural transmission of the behaviors she fostered, garnered worldwide attention. This was one of the first examples of a newly introduced behavior's spreading through a population of animals. In addition, primates are clever, and so although many didn't believe we would ever find evidence of culture in animals, we could be somewhat comforted that the first strong evidence came from a smart animal. But perhaps just as important, the new behavior—washing food—seemed eerily human-like. It doesn't take much imagination to picture our human ancestors learning the same trait, in a similar manner, hundreds of thousands of years ago.

Imo's actions were not the only ones to garner attention, as many similar cases are now on record. Michael Huffmann, for example, has found another case of primate culture that stirs pseudo-nostalgia for our human ancestors who were trying to scratch out an existence during prehistoric times. Many of the movies, and some of the science, associated with prehistoric humans centers on how we used stones. Stones served as weapons in combat and as a means to take down dangerous prey. In all likelihood, stones also served ritual functions as well. But stone use, and the cultural transmission of stone use, are not uniquely human.

Huffmann's study revolves around twenty years of work on the Japanese macaques of the Iwatayama National Park in Kyoto. Early on in his work, he began to observe a behavior never before noted in macaques: individuals would play with stones, particularly right after eating. This bizarre behavior began in 1979 when the awkwardly named Glance-6476, a three-year-old female, brought rocks in from the forest and started stacking them up and knocking them down. Not only that, but Glance-6476 was very territorial about her stones and took them away when other monkeys approached. When Huffman returned to Glance-6476's troop four years later, stone play (often referred to as stone handling) "had already become a daily occurrence [and] was already being transmitted from older to younger individuals." Interestingly, cultural transmission in this system seems to work down the age ladder, but not up. Although many individuals younger than Glance-6476 acquired her stone play habit, no one over her age added this behavior to their repertoire.

The reason for opening a discussion of nonmating examples of cultural transmission with the monkeys of Koshima Islet and the Iwatayama National Park is not that they are necessarily the most convincing cases of cultural transmission recorded in animals. These examples lack the controls that many other studies of cultural transmission contain. But food washing and stone handling do dramatically illustrate that culture can be a powerful force outside the context of mate choice.

If it turned out that cultural transmission was important in mate choice alone, biologists would be surprised. But such a finding, in and of itself, would not force us to rethink the role of culture in animal life in general, or the implications of animal work on culture for understanding human social evolution. In fact, culture has been shown to be a powerful force in shaping animal behavior in every context, from learning who the enemy is to avoiding ingesting dangerous, untested new foods.

THE SWEET SMELL OF SUCCESS

Few people have worked longer or harder at trying to unite biologists and psychologists interested in cultural transmission than Bennett Galef. For the past twenty-five years Galef has been involved in some of the most interesting work yet undertaken on social learning and food gathering. His species of choice, the rat, may not be high on most people's list of charismatic animals, but it is ideal for studying how animals use cultural transmission to determine what they should and shouldn't add to their diets.

When it comes to feeding, rats eat what other species won't. Being scavengers, they are presented with opportunities to sample new foods all the time. This probably has been true for most of the rat's long evolutionary history, but it has been particularly the case over the last few thousand years, when humans and rats have had a close, if not pleasant, relationship. Many rat populations use discarded human food remains as a primary source of nutrition, which means that when humans add new items to their diet, so may rats. And therein lies a dilemma. One the one hand, a new food source may be an unexpected rat bounty. On the other hand, new foods may be dangerous, either because they contain elements inherently bad for rats, or because one doesn't know how a new food should smell, and so it is difficult to tell if something you come upon is fresh or spoiled. This is the ideal environment for social learning to take hold, and take hold it surely does.

The effect of social learning and food preference in rats runs deep—and that is putting it mildly. Although I shall focus on Galef's studies on juvenile and adult rats, imitation-like factors operate at much earlier stages in a rat's life. In fact, it all starts in the womb. Rat fetuses can actually sense what type of food their mother is eating late in pregnancy, and prefer that food themselves shortly after they are born. It's worth noting that the food used in the experiment demonstrating this ability was cloves of garlic, and chances are that rats have had little exposure to this delicacy during their evolutionary history. What that means is that the finding that mom's food preference has in utero effects on junior's subsequent choice of foods wasn't based on some genetic similarity between mother and offspring in terms of innate food preferences (since no such preferences could exist for garlic). Once junior is born, the same sort of thing is found with respect to chemicals in mom's milk. If milk contains the flavor of some particular food, infant rats quickly acquire a preference for that food.

The story of social learning and food preferences in rats started out with a test of what is known as the information-center hypothesis. The idea here is that in species where environmental cues about food are constantly shifting, foragers may learn critical tidbits about the location and identity of food by interacting with others that have recently returned from a bout of food hunting. Galef and his colleague Stephen Wigmore tested this hypothesis in a species that finds itself in exactly the situation described by the information-center hypothesis: the Norway rat. To test whether social learning played a role in rat foraging, rats were divided into two groups: subjects and demonstrators. The critical question was whether subjects could learn about a new, distant food source simply by interacting with a demonstrator that had experienced such a new addition to its diet.

After living together in the same cage for a few days, the demonstrator rat was removed and taken to another experimental room, where it was given one of two new diets: either rat chow flavored with Hershey's cocoa or rat chow mixed with ground cinnamon. The demonstrator was then brought back to its home cage and allowed to interact with the subject for fifteen minutes; then the demonstrator was removed from the cage. For the next two plus days, the subject rat was given two food bowls: one with rat chow and cocoa, the other with chow and cinnamon. Keep in mind, again, that these rats had no personal experience with either of the novel food mixes they were experiencing, nor did they ever see their demonstrator eat these new food items. The results of this experiment were clear. Subject rats were influenced by the food their demonstrators had eaten in a far-off room, and they were more likely to eat the food that their demonstrators ate.

Rats are primarily olfactory-oriented creatures, and so it isn't surprising that the way that they learn what demonstrators had for their last meal is by smell. Galef and Wigmore used this reliance on smell to follow up their initial work with an experiment designed to see how long demonstrators remained useful sources of information about new food. In addition to placing demonstrators with subjects directly after demonstrators ate their novel food items, a delay of thirty and sixty minutes was also examined. The thirty-minute case yielded the same results obtained in the original work, but an hour seemed to be long enough for the system to crash, with subjects no longer likely to mimic the choice of their demonstrators. Whether the scent simply dissipated during the longer delay or the rats ignored it after the longer time frame is hard to say.

Once rats use their peers to learn what food sources out in the environment are worth trying, they hold on to this information with a vengeance—so much so that they will even override their own experience with what is good to eat and what isn't based on what they learn from others. For example, give rats food laced with cayenne pepper, and they will normally have nothing to do with it. Train a rat to interact for long periods with another rat that has been fed cayenne pepper, however, and all of a sudden it adds cayenne-laced food to its diet. Of course, one might rightly say that this is the sort of thing that would backfire on rats only when they are manipulated by a devious researcher in the lab. In nature, rats would rarely have long-term interactions with another individual who was constantly choosing a food that they found repulsive. Nonetheless, the power of this experimental work lies not in its tight relation to the natural world per se, but rather in what it says about the power of social learning in shaping general foraging behavior.

As was hinted in the cayenne pepper experiment, there is an interesting, unexpected twist in the rat social learning story. Given how strong a force social learning is in determining what a rat should eat, one might expect that it also plays a significant role in helping a rat determine what to avoid eating. This apparently is not the case. Galef and his colleagues injected some unlucky rats with lithium chloride and used such individuals as demonstrators. The demonstrators were obviously ill, and yet observers were perfectly content to add on the new food item emanating from their ill cagemate. The fact that the new food item they smelled on their demonstrator was likely the cause of that individual's illness seemed to go by unnoticed. A similar (but not nearly as detailed) story can be told of culture and foraging in the rat's dreaded enemy, the cat.

Why social learning is so finely tuned in helping rats choose what to eat, but not what to avoid, remains something of a mystery.

THE SWEET SIGHT OF SUCCESS

As with mate choice, if cultural transmission played a role in foraging in only a single group of rats, it might be of some academic interest, but it would hardly be the sort of stuff that helps frame a new view of how behavior spreads through populations. Fortunately, cultural transmission in the context of foraging has been purportedly recorded in many species, including cats, dolphins, lions, chimpanzees, coyotes, squirrels, moose, otters, meerkats, crows, and whales. The best evidence can be seen in studies on birds and humans, for unfortunately much of the other evidence was not gathered in methodologically sound ways.

In the 1940s tea-drinking Brits were becoming increasing irritated that the foil caps on top of their milk bottles were being torn off before they could retrieve the freshly delivered bottles from their doorsteps. The bottles were being tampered with by *Parus caeruleus,* or blue tits, as they are known. J. Fisher and Robert Hinde put forth the notion that this new behavior had been accidentally stumbled on by a lucky blue tit and that others learned this nifty trick, at least in part, from watching the original milk thief(s). In addition to ruining a lot of milk, it was reported that blue tits were using their new skills to tear down wallpaper. The mischief seemed to be breaking out simultaneously all over the United Kingdom. The consensus was that this was a clear case of cultural transmission of a newly acquired feeding habit. But as charming as this story is, and as much as it paints a fair picture of the power of culture to shape animal behavior, the work is neither experimental nor controlled. David Sherry and Galef (of imitation and rats fame), for example, show that chickadee birds can be trained to open bottles by simply observing another bird in an area that will eventually have a milk bottle, even when the bird observed is not seen in the presence of a milk bottle. Of course, this doesn't mean that the birds in the original study fifty years ago didn't use cultural transmission, only that they didn't necessarily use it. This is precisely the sort of thing that happens in uncontrolled work on evolution and behavior. For the most comprehensive, controlled work on foraging and cultural transmission, we must turn to the avian equivalent of the rat: the pigeon *(Columbia livia)*.

Like rats, pigeons are an ideal species in which to examine cultural transmission of feeding behavior. Being primarily scavengers feeding on human garbage, pigeon diets change with the fanciful whims of the human taste bud. This uncertainty in what new foods will be available and which are safe favors the transmission of behavior via paths such as imitation. Over the past fifteen years, Louis LeFebrve and his colleagues Luc-Alain Giraldeau and Boris Palameta have run an intriguing series of experiments that attest to the strength of cultural transmission in shaping diet in the pesky, but malleable, pigeon. This work has focused on three related issues: (1) What type of information do pigeons transfer about food? (2) How does such information spread, or fail to spread, through a population of pigeons? and (3) What factors favor the cultural transmission of information over alternative means of acquiring information?

When Palameta and LeFebrve did their first studies on copying and foraging in pigeons in 1985, they had their work cut out for them. Although cultural transmission as a means of information transfer was accepted in primates, controlled work on social learning and foraging in birds was virtually nonexistent. In fact, with a few exceptions (including Galef's work on rats), controlled work on cultural transmission of food preferences in nonprimates was almost impossible to find. Furthermore, when such studies were undertaken, procedural problems usually made it impossible to distinguish cultural transmission from alternative explanations such as "blind copying" and "social facilitation." In blind copying, simply seeing an individual in a novel situation elicits a behavior that the observer already knew; in social facilitation, individuals are drawn to an area by seeing another animal there and then learn the new task on their own.

Palameta and LeFebvre set out to examine cultural transmission in a three-part experiment that used observer and demonstrator animals. The task that observer pigeons needed to master was piercing the red half of a red and black piece of paper covering a box. Under the paper was a bonanza of seeds for the lucky bird that made it that far.

An observer pigeon was placed in an arena with such a

food box (half red and half black) and exposed to one of four scenarios. Some unlucky birds saw no model on the other side of a clear partition. None of the pigeons in this group learned how to get at the hidden food. In the blind imitation group, observers were shown a model that was eating from a hole in the paper. The hole was made by Palameta and LeFebrve, and as such, although observers did see a model eating, they did not see the model solve the hidden food puzzle. Once again, pigeons in this treatment failed to learn how to get food from the multicolored box. In the final two treatments of the experiment, birds saw either a model pierce the red side of the paper but get no food (the "local enhancement" case) or a model both pierce and eat. Although birds in both of these treatments learned to solve the food-finding dilemma, those in the latter treatment did so much faster.

Given all the examples we have touched on, culture and pigeon foraging might not seem particularly exciting. Keep in mind, however, that at the time this work was done, there were very few controlled experiments like this on the books, and none focused on bird foraging and culture.

There is an interesting twist to the pigeon story. In many group-living animals, when it comes to foraging, there are two strategies individuals use: producing and scrounging. Producers find and procure food, while scroungers make their living parasitizing the food that producers have uncovered. Layered on to the imitation and foraging story we have seen in pigeons, we find producers and scroungers in this species as well. And it is the unusual way that producing and scrounging interacts with imitation that makes the pigeon story particularly useful in furthering our understanding of cultural transmission.

Despite Palameta and LeFebrve's work demonstrating cultural transmission in pigeons in cages, when birds are tested in groups, only a few birds learn new feeding behaviors by observing others. This apparent paradox caused Giraldeau and LeFebrve to examine whether scrounging behavior somehow inhibited cultural transmission. To do so, they used a different set-up than described above. Now, flocks of pigeons were allowed to feed together. Forty-eight little test tubes were placed in a row, and five of these tubes had food. Which five, of course, was unknown to the birds. To open a tube, an individual had to learn to peck at a stick in a rubber stopper at the top of the tube. This caused the test tube to open and the contents to spread over the floor below. Once the food was out, any bird in the vicinity, not just the one that opened the tube, could eat it.

Results of their experiment were striking. As Giraldeau and LeFebrve predicted based on earlier work, only two of the sixteen pigeons in their group learned to open tubes. That meant that the flock was composed of two producers and fourteen scroungers. Two additional findings suggested to Giraldeau and LeFebrve that scrounging inhibited learning how to open tubes via observation. To begin with, scroungers followed producers and seemed more interested in where producers were than what producers did to get food. Second, by removing the two producers from the group, Giraldeau and LeFebrve were able to ascertain that scroungers not only didn't display the tube-opening trait, but they also didn't know how to open these tubes. That is to say, it was not as if scroungers could open the tubes but opted not to; they truly never learned the trait when producers were around.

In order to get a firm handle on just how scroungers blocked cultural transmission of foraging skills, Giraldeau and LeFebrve ran a second set of more controlled experiments. Here, a single observer was paired with a single demonstrator (that already knew how to obtain food). If an observer had the chance to view a demonstrator open tubes and obtain food, in time the observer learned how to open tubes. That is, all birds were capable of learning the foraging task. In an ingenious manipulation, Giraldeau and LeFebrve then set up the experimental cages so that every time the demonstrator opened a tube, the food in that tube slid over to the observer's side of the cage. In these treatments, the observers rarely learned how to open the tubes themselves. Their scrounging on the food that others had found interfered with their ability to learn from others. Giraldeau and LeFebrve's finding remains one of only a few cases in which researchers knew enough about a specific case of cultural transmission to figure out when it is effective and when it isn't.

We humans learn almost everything important about our diets from cultural transmission. Consider this statement by Paul Rozin, the dean of diet and cultural transmission in *Homo sapiens:*

> For Hindu Indians, food is a moral substance. It is not an impersonal nutrient package, bought in a supermarket, but the bearer of the mark of its particular maker. The food one eats has something like an essence of its preparer. One can harm one's status or persona by consuming food prepared by a person of inferior rank. Indeed, one can reconstruct the elaborate caste structure of India from examining who can eat whose food.

Such a strict caste structure is rarely dictated by national cultures now, but how could cultural transmission ever have such a dramatic impact on diet in humans? Don't we just eat whatever we feel like?

In looking at the types of experiments that are undertaken when tackling this question, the inherent difficulties faced in studying cultural transmission and food preferences in humans become obvious. Back in 1938, K. Duncker examined the food preferences of preschoolers as young as two and a half years old. After establishing what a child liked and disliked, that child was given the opportunity to view another individual choose one of the foods that he himself did not rank particularly highly. Although the effects that Duncker uncovered lasted only a week or so, he did in fact find that children shifted their food preference toward that of the individual they observed. The effect was especially strong just where one might expect: when the child was observing a friend or a powerful child who wasn't a friend. In addition, Duncker read stories to children in which a hero figure ate the food that the child ranked low, and sure enough, that food took on new meaning to the child subject.

In a similar but more controlled vein, L. Birch ran a fascinating experiment on children's acquisition of food preferences some forty years after Duncker. Birch began by recording the vegetable preferences of children before the start of a trial. Such children were then placed at a table where three or four other youngsters chose what vegetable they would eat before the child of interest (the observer). Slowly but surely, subjects began to shift their choice toward that of their peers. Given that none of the models in either Birch or Duncker's work were actively teaching the subjects (e.g., they weren't explaining why their choice was a good one), these studies show how strong cultural forces can be, at least with respect to children—so strong that foods initially producing a strong negative effect (such as chili pepper) can eventually be enculturated into a child's diet.

The question then naturally arises as to why we rely so heavily on this mode of information transfer when it comes to diet. The answer probably lies in the fact that the foods that we eat, particularly meat dishes, are breeding grounds for bacteria and other disease-causing agents, the prevalence of which varies at a rate far too quick for human genetic responses possibly to cope with. If we couldn't culturally learn what was good and what wasn't good to eat, we'd be in serious trouble. Biologists have mused about renaming our species *Homo imitatus.*

Note that work on the import of imitation in humans begins with children even younger than described above. Consider the 1993 field experiment undertaken by E. Hanna and A. Meltzoff on fourteen-month-old infants. Some infants—the tutors—were taught to play with a toy in a novel fashion. These tutors were then brought to a series of day care centers none of which they had ever visited before. Other infants "sat around a table, drinking juice, sucking their thumbs and generally acting in a baby-like manner," while the tutor played with the toy in a novel manner. Two days later, the observer babies were examined in their own houses (not the day care center), and it was obvious that they had adopted the novel toy-playing behavior. Consider that the next time someone tells you that television doesn't affect your child's behavior.

GOOD ROLE MODELS, BAD ROLE MODELS

Up to this point, most of the examples of cultural transmission in animals that we have touched on have one interesting similarity: all the information transferred was useful in one way or another. Be it information on getting a mate or learning what to eat or not to eat, so far the information transferred in a bout of cultural transmission has been something an animal can use to increase its fitness. But need that be the case? Couldn't maladaptive as well as adaptive information change hands by cultural transmission in animals (as we have already seen is the case for humans)? Enter Kevin Laland, Kerry Williams, and the guppy.

Laland, who has studied with many of the premier evolutionary biologists interested in culture, and Williams wanted the answer to the following question: If imitation is a quick and efficient way of transmitting information across a population, can this ever result in a maladaptive behavior spreading through groups? That is, can mistakes become the norm if enough individuals copy each other? To answer this question, they devised a James Bond–like fish tank, trap doors and all, and created two paths to a single food source. One path was short and efficient and the other longer and more circuitous.

Before starting the cultural transmission phase of the experiment, single guppies were tested in a fish tank. Presented with the choice between the long and short paths, the guppies consistently learned to take the shorter route. Given what we have learned about guppies, this shouldn't surprise you all that much. But what came next just might. After establishing that guppies were smart enough to learn the shortest distance to a food source, Laland and Williams trained two groups of fish. Fish in the first group were easily trained to take the short path to food. Training in the second group was a bit more devious; it involved shutting trap doors, so the fish in this group were frightened away from the short route. The second group, then, was trained to take the longer path.

Once the "long path" and "short path" groups were trained, Laland and Williams, over the course of time, slowly removed the original group members in each group and replaced them with new individuals. So initially a group of five guppies was made up of five trained individuals, then four trained individuals and one untrained, then three trained individuals and two untrained, until at the completion of the experiment none of the original trained group members remained. The question is whether the fish remaining at the end of the experiment—none trained to a particular path—maintained the "traditional" path taken by fish that were originally in their group.

Laland and Williams discovered that the short-path group used that path even when none of the trained individuals remained in the group. That is, social learning allowed the transfer of useful information, even when the number of trained individuals decreased over time. The same result was uncovered for the long-path groups. At the end of the experiment, group members in the long-path treatment still took the long path, although none of the original models that had taught that maladaptive behavior were around. So social learning can lead to both efficiency and mistakes. Perhaps the most interesting finding was that fish trained in the long-path group actually took longer to then learn the short path than did individuals with no experience. Social learning in the long-path groups not only resulted in guppies' acquiring the wrong information, it actually made it more difficult to learn subsequently what was the optimal path to take. Cultural transmission in guppies, for good or bad, clearly is a force to be reckoned with. And cultural information as a means of navigating around unknown spaces is not restricted to guppies.

Back in 1984, Gene Helfman and Eric Schultz ran an experiment similar to that undertaken by Laland, except their work was undertaken in the wild (hence, lacking some of the controls Laland's lab work allowed) and focused exclusively on the benefits of cultural transmission. Helfman and Schultz studied the french grunt, a tropical coral reef fish. Juvenile grunts form groups that are loyal to a particular site on the coral reef for upward of three years. Each night, groups of fish migrate from their favored location and use what appear to be featureless paths that lead them to nearby nocturnal feeding grounds. Daybreak comes, and the fish migrate back to the reef.

Helfman and Schultz wished to know whether fish learned the migration path through observing others. In order to tackle migration and cultural transmission, they tested two categories of juvenile grunts. One group was moved from their home location to a new coral reef spot, stock full of resident fish, and the second group was moved to a new coral reef location, but only after all the residents were removed. When examining whether transplanted fish learned the migration route that was associated with their new home, they found that only those that had the opportunity to observe resident fish make the trek were able to learn new migration routes. Fish that had no one to learn from were lost when it came to where to feed at night. They basically tried to use the route they took at their old home and hence were doomed to fail.

LEARNING WHO THE BAD GUYS ARE

Individuals belonging to the tropical species of bird known as motmots instinctively fear poisonous coral snakes. The particular coral snakes that are dangerous to motmots have a specific color pattern: red and yellow bands. Present baby motmot chicks with a wooden dowel with red and yellow bands painted on it, and the chicks instantly fear it. Paint green and blue bands or even red and yellow stripes on the dowel (neither of which resemble snakes dangerous to the motmot), and the motmot youngsters peck at it and do other things that would be very dangerous if it were a snake. From birth on, motmots know who the enemy is.

The motmot solution to knowing who the enemy is—hardwiring the answer into the genetic code—is an eminently reasonable one, but it works only under certain conditions. If there are lots of different predators to handle or if the identity of predators is constantly changing, innate fears may be an inadequate or inappropriate solution to the "know your enemy" problem. Under such conditions, it might pay to learn who the enemy is, and the best way to do that might just be to observe how others respond to potential threats. Eberhard Curio and his colleagues did just that.

Blackbirds, like many other bird species, undertake a fascinating antipredator defense mechanism called "mobbing." Mobbing is truly a bizarre sight to witness. Once a flock of blackbirds spot a predator, some of them join forces, fly toward the danger, and aggressively attempt to chase it away. Such united fronts often work well enough to force predators to leave the blackbirds' area. Exactly how mobbing works and how one calculates what the benefits and costs for such an action may be are still a matter of debate, and many cost-benefit models on this subject can be found in the avian behavior literature.

As we learned in Chapter 5, Curio and his colleagues examined whether one function of mobbing behavior may be to help novice blackbirds identify predators. This is referred to in the literature as the "cultural transmission" hypothesis, and it "suggests that perceiving other birds mob an object may teach an individual to fear that object and thus subsequently avoid it, mob it more strongly or both."

Here's how Curio's study worked. Every trial started with a "teacher" and a "naive" bird, each in its own aviary. The experimental apparatus was designed so that each of the birds could see another bird—this time a noisy friarbird. The friarbird was a novel predator that neither the teacher nor the naive bird had seen before. Furthermore, it looked nothing like any known predator of blackbirds. The catch was that the friarbird was presented in such a way that the naive bird saw it, but the teacher saw both a friarbird and, adjacent to it, a predator of blackbirds (a little owl, *Athena nocuta*). From the viewpoint of the naive subject, the little owl was out of sight. What that translates into is a situation where the teacher mobs the little owl and the naive bird sees it mobbing a friarbird. With this, it is possible to test whether mobbing teaches naive blackbirds that the friarbird is indeed a predator.

Sure enough, Curio and his collaborators found that once they had seen a teacher apparently mobbing a friarbird, naive birds themselves were much more likely to mob this odd new creature than if they had not been exposed to a teacher. Curio took the experiment one step further and essentially asked whether the subject (now not so naive) could then act as a teacher for a new naive bird. And if that worked, how many times could he get a former naive bird to act as a teacher? In other words, how long is the "cultural transmission chain" in blackbirds? The longer such chains are, the more powerful cultural transmission may be in spreading novel behaviors through a population. Although his sample was small, Curio found that the blackbird cultural transmission chain was at least six birds long. Not bad for the mind of a bird.

Most primatologists since the early 1900s have realized that many monkey species have a strong fear response in the presence of snakes. The question is whether it is an innate response, a learned response, a response acquired culturally, or some combination. Early work comparing wild- and laboratory-raised primates suggested that fear of snakes per se was not completely innate, as laboratory-raised individuals (that never saw a snake) did not respond to snakes in the same manner as wild-raised individuals (that had the chance to experience snakes in the wild).

In controlled laboratory experiments with juvenile rhesus monkeys, Susan Mineka and her colleagues attempted to test whether cultural transmission played a role in the fear of snakes. She began with juvenile rhesus monkeys unafraid of snakes. These monkeys quickly adopted a fear response after watching a model respond to snakes with typical fear gestures and actions. Interestingly, it made no difference whether the individuals they watched were their parents or unrelated monkeys. Regardless of blood ties to the model, the newly snake-fearing juveniles retained their culturally derived fear for at least three months.

ANIMAL MODELS VERSUS ANIMAL TEACHERS

It is one thing to claim that animals transfer information by cultural transmission, but quite another to say that animals teach each other. Many kinds of cultural transmission, such as imitation, can be passive in the following sense. Observers may be gleaning new information from others, and incorporating what they learn into their repertoires, but nothing special is going on from the perspective of the individual being copied. That is, the individual being copied needn't be doing anything besides what it always does. It just so happens someone is observing and then imitating.

Teaching is another can of worms. Teaching implies that someone serves as an instructor—a much more active, not to mention complicated, role than simply being a guide that someone copies (as in imitation). And not surprisingly, the notion that animals teach each other is perhaps the most continuous aspect of animal cultural transmission. For if animals do teach, then another of the behaviors that we've viewed as uniquely human bites the dust. We still might be the only animals that teach morality, but we'd no longer be the only animals that teach period, and for some that might be hard to swallow.

Tim Caro and Marc Hauser tackle the question of animal teaching head on in a fascinating 1993 review paper. While many definitions (both scientific and colloquial) of teaching exist, Caro and Hauser argue that one component of such definitions is unnecessary. Most definitions assume that the teacher must have some knowledge of the pupil's mental state. Caro and Hauser disagree and argue that while some forms of teaching include such abilities on the part of teachers, they are not needed for teaching to have taken place. Instead they opt for a less cognitively based definition:

> An individual actor A can be said to teach if it modifies its behavior only in the presence of a naive observer B, at some cost or at least without obtaining an immediate benefit for itself. A's behavior thereby encourages or punishes B's behavior, or provides B with experience, or sets an example for B. As a result, B acquires knowledge or learns a skill earlier in life or more rapidly or efficiently than it might otherwise do, or that it would not learn at all.

So essentially, to teach a naive individual, what one does must not provide an immediate benefit, must be done for naive "students" only, and must impart some new information to students faster than they would otherwise receive it. Now, we can always quibble with different aspects of this (and Caro and Hauser themselves go into more detail on every component of their own definition), but it might be more illuminating in this case just to see how common teaching is, even under Caro and Hauser's definition. A list of some examples of putative teaching include the following:

- Mothers changing their hunting behavior when their young are old enough to start learning how to hunt. As an extreme example, consider a female cat that captures live prey and allows her young to play with prey, making sure it doesn't escape. This sort of teaching has been documented to differing degrees in domestic cats, lions, tigers, cheetahs (particularly strong evidence here), meerkats, mongooses, and otters.

- In chimpanzees, gorillas, rhesus monkeys, yellow baboons, and spider monkeys, "mothers have been observed encouraging their young to walk and follow them, typically in the context of group movement or foraging."
- Adult chacma baboons and squirrel monkeys chase juveniles away from objects that the adults know are dangerous.
- Washoe the chimpanzee, an expert in sign language, may have taught Loulis (a less adept signer) the correct way to sign for "food."
- Vervets may teach their young the appropriate alarm call for different predators.
- Chimpanzees in the Tai National Forest of the Ivory Coast may teach their young how to crack open nuts. For example, some females leave a cache of nuts on top of an anvil in the vicinity of young that are approximately the age where they begin to learn nut-cracking skills.
- A number of species of raptor birds (such as hawks, falcons, and ospreys) appear to teach their young the difficult art of precise hunting while flying at high speeds.

In their search for common themes that underlie these possible cases of teaching, Caro and Hauser uncovered two possible candidates. First, virtually all cases involved parents teaching young individuals. At first this might not seem surprising, but if you consider that young individuals can learn from others besides their parents, and adults could presumably teach other adults, the parent/teacher–offspring/student relationship suggests something about the costs and benefits of teaching. The kinship that bonds teacher and student may be the only benefit large enough to make up for the costs of teaching. Second, Caro and Hauser found two kinds of teaching, which they label "opportunity teaching" and "coaching." In the former, teachers put students in a "situation conducive to learning a new skill or acquiring knowledge," and, in the latter, the teacher "directly alters the behavior . . . by encouragement or punishment." The majority of examples of animal teaching fall under opportunity teaching—not surprising, given that opportunity teaching is arguably the simpler of the two types of teaching.

A MATTER OF LIFE AND DEATH

None of the studies reviewed thus far in this chapter shows a direct, measurable, clear impact on lifetime survival. The reason, of course, is that it is extraordinarily difficult to isolate any single behavior and examine its contribution to survival. Yet that is the stuff on which natural selection is built, and so when those sorts of data are available, it is always quite exciting, and when they are available, with respect to the cultural transmission of some behavioral trait, it is newsworthy. My former graduate student, Michael Alfieri, has collected just such data in guppies.

Alfieri's Ph.D. dissertation work examined the effect of

learning (in this case, learning about how to escape predators) on survival. To begin, he exposed guppies to various "antipredator learning treatments" to examine the effect of learning on direct survival. Male guppies would, for example, watch another guppy undertake an antipredator behavior but survive, or see another guppy chased and finally eaten by a predator, or simply see a guppy dropped into the mouth of a predator. What he found was that the type of information that an observer obtained had a significant impact on its survival. Males that saw another guppy chased and finally eaten by a predator were much less likely to be eaten by a predator themselves in subsequent tests. Such individuals presumably learned what works and what doesn't work in terms of getting away from predators. It might surprise you to know that this is one of the only experiments on the books to show a direct, large, and uncontroversial effect of learning on fitness. The beauty of the system Alfieri developed is that the finding just described is not the most exciting thing he uncovered.

After establishing learning and survival rates, Alfieri took a group of predator-naive fish and split them up. In one group, each naive fish would be paired up with another fish that itself had learned appropriate antipredator behaviors. The other group of naive fish would be paired with someone like themselves—someone with no predator experience. What he found was that fish that were paired with experienced partners survived interactions with a predator with the same probability as their partner; naive fish paired with inexperienced partners suffered much higher mortality rates. So not only does personal experience have a direct effect on survival in the face of a predator, but the experience level of your partner does as well. Exactly how naive fish benefit from their partner's experience is not yet known. Nonetheless, by any reasonable definition, what Alfieri found was an immediate fitness benefit associated with cultural transmission, and this is just the type of data that we so desperately need to develop a complete theory on how cultural transmission truly affects animal behavior.

THE POWER OF CULTURE

The work outlined in this chapter shows that cultural transmission as a force in nature is far from restricted to the arena of mate choice. The more general the work on culture in animals is, the more critical an issue it is in our understanding of why organisms, including ourselves, act the way that we do. If cultural transmission is restricted to mating, then biologists interested in mating need to pay attention to this force (and for the most part, they have not). But if feeding, protection, communication, navigation, and friendship are affected by cultural transmission in animals, then in principle, no behavior is off limits in terms of culture.

The vast array of animal behaviors that are touched by the long fingers of culture continues to grow, and my guess is that we have seen only the tip of the iceberg. Once behavioral ecologists stop giving cultural transmission short shrift, there is no telling the number of examples of animal culture that we will undercover. More important, the general lessons we may learn from understanding the impact of cultural transmission as a force in nature are just beginning to surface.

Once researchers give cultural transmission the time it merits, our general view of why things behave as they do may change dramatically. The idea that a bad role model can spread bad behavior through a large human population is familiar to readers of sports journalism. But experimentally studying this phenomenon, as psychologists report, is difficult. That being the case, the kind of detailed work undertaken by Laland and Williams on guppies is likely to guide future studies of information transfer through human populations.

No single animal study could explain human behavior, because every animal makes and breaks its own cultural rules in its own way. That being said, Laland and Williams's work on how education really works, and the many other studies outlined in this book, serve as a starting point for controlled experiments looking for common themes that underlie animal culture. Arriving at a general theory of cultural transmission, and its implications for understanding behavior, is just a matter of time.

AFTERWORD

Understanding Our Behavior

> Errors and exaggerations do not matter. What matters is boldness in thinking . . . ; in having the temerity to proclaim what one believes to be true without fear of consequences. If one were to await the possession of the absolute truth, one must be either a fool or mute.
>
> José Clemente Orozco

THE MOST VEXING QUESTION SCIENTISTS have ever tackled is also one of the most familiar questions in everyday life: Why do people behave the way they behave? Despite huge advances during the last century in biology, psychology, and anthropology, this question is still as mysterious as ever. To begin to comprehend the nature of behavior requires a thorough understanding of the process of cultural evolution, because we have evolved to act in terms of others' actions, usually by simply mimicking them. Of course, we *Homo sapiens* make choices about whom to mimic, and our replications of others' behavior usually begin with some creative original act, but these are not prerequisites to culture. Every individual human being replicates the actions of many others. Cultural evolution has had a profound impact on the social fabric of animal life, and its effects will only increase with the passage of time. Animals may not choose whom to mimic, and those chosen may not be doing anything particularly creative, and yet cultural transmission can still be a powerful force.

The zoological work on cultural evolution reveals strange and even amazing facts about animals no matter how large or small their brains are—indeed, some just barely have what we can call a brain. The actions of a few individuals, or even a single one, can dramatically shift the evolutionary future of a particular population fundamentally because individuals are keen copiers. If a few individuals all of a sudden prefer behavior X rather than Y and others copy this, our population is now full of Xers and may stay that way. Or behavior Z might pop up on the scene, and depending upon who does it and who watches, Z might become more and more common. And remember, these shifts are not necessarily due to one behavioral strategy's being more fit (in genetic terms) than another. An individual did something original, and it simply became fashionable. That is a dramatic break with standard genetic theory.

On the other hand, genes *were* first. For most of evolutionary history, life has been ruled by genes. Exactly when cultural evolution began to run away and outpace genetic evolution is hard to say, because in each species, in each population, the pace of each kind of evolution proceeds at a different rate. But simply because we don't know when cultural evolution took over, and because of the fact that genes existed first, should we spend all our scientific research funds on investigating genes and virtually none on the science of cultural evolution?

Orders of magnitude separate the numbers of scientists sequencing genes from those studying the evolution of culture. Could there be an equivalent of the Human Genome Project for cultural evolution? Right now it would be premature because there is still no unified theory of the biology of culture. But just as sequencing genes will have a profound impact on human life, so too can understanding the roots of cultural behavior. The more we understand the nature of why animals do what animals do, the more we understand ourselves. We can't predict specific results immediately from that enterprise, but we can be certain it will be profoundly valuable.

The implications of a comprehensive theory of cultural evolution will be tremendous. I believe that the establishment of such a scientific theory will be no less profound in the history of our species than the landmarks of achievement we have seen in physics in the last century, such as Einstein's theory of relativity and quantum field theory. Indeed, perhaps the achievements in the science of matter in the twentieth century will even be overwhelmed by achievements in the science of life in the twenty-first century. *Homo imitatus* will ensure a long life for the scientific insights we achieve.

Selected Additional Readings

C. Darwin. 1859. *On the Origin of Species.* London: J. Murray.

C. Darwin. 1871. *The Descent of Man and Selection in Relation to Sex*. London: J. Murray.

These two classic works outline Darwin's theory of evolution by natural selection. *The Descent of Man and Selection in Relation to Sex,* as its title implies, focuses on how natural selection operates on mate choice.

G. J. Romanes. 1895. *Mental Evolution in Animals.* New York: Appleton.

G. J. Romanes. 1898. *Animal Intelligence.* 7th ed. London: Kegan Paul, Trench, Trubner and Co.

Romanes, a psychologist by training and a good friend of Darwin, presents his theory on imitation and intelligence in animals. These books are often viewed as initiating the field of evolutionary social learning.

E. O. Wilson. 1975. *Sociobiology: The New Synthesis.* Cambridge: Harvard University Press.

C. Lumsden and E. O. Wilson. 1981. *Genes, Mind and Culture.* Cambridge: Harvard University Press.

C. Lumsden and E. O. Wilson. 1983. *Promethean Fire.* Cambridge: Harvard University Press.

In 1975, E. O. Wilson presented his unifying theory of behavioral evolution, applying "natural selection thinking" to behavioral traits. Lumsden and Wilson's follow-up books to *Sociobiology* explore how the new field of sociobiology incorporates the notion of nongenetic transmission of information.

L. L. Cavalli-Sforza and M. W. Feldman. 1981. *Cultural Transmission and Evolution: A Quantitative Approach.* Princeton, N.J.: Princeton University Press.

This very mathematical treatment of cultural evolution, written by two world-class population geneticists, is the first book-length treatment of the evolutionary models underlying some forms of cultural transmission.

R. Boyd and P. J. Richerson. 1985. *Culture and the Evolutionary Process.* Chicago: University of Chicago Press.

Boyd and Richerson's book on cultural evolution is the most widely cited monograph devoted to conceptual and mathematical issues surrounding cultural transmission.

T. R. Zentall and B. G. Galef, eds. 1988. *Social Learning: Psychological and Biological Perspectives.* Hillsdale, N.J.: Erlbaum.

C. M. Heyes and B. G. Galef, eds. 1996. *Social Learning in Animals: The Roots of Culture.* London: Academic Press.

This pair of edited volumes provides a thorough, cross-species, cross-disciplinary overview of studies on imitation and social learning in animals.

M. Andersson. 1994. *Sexual Selection.* Princeton, N.J.: Princeton University Press.

The bible of sexual selection studies undertaken through 1994. With more than two thousand citations, this book provides a remarkable overview of sexual selection.

S. Blackmore. 1999. *The Meme Machine.* Oxford: Oxford University Press.

Blackmore provides a book-length treatment of memes—the cultural units of transmission first created by Richard Dawkins in *The Selfish Gene* (Oxford: Oxford University Press, 1976).

Notes

CHAPTER 1: THE CULTURED ANIMAL

7 *Gregor Mendel, surely the most famous Austrian monk:* Mendel's work was independently rediscovered by Erich von Tschermark, Carl Correns, and Hugo De Vries. E. Minkoff, 1983, *Evolutionary Biology,* Reading, Mass.: Addison-Wesley.

7 *In fact, the word* gene: In 1909, Danish geneticist W. L. Johannsen introduced the word *gene* (by shortening *pangen*) and noted it was to be thought of as "a kind of accounting or calculating unit." W. Johannsen, 1909, *Elemente der Exakten Erblichkeitslehre,* Jena: Gustav Fisher; E. Mayr, 1982, *The Growth of Biological Thought,* Cambridge, Mass.: Harvard University Press.

8 *Darwin's ideas about how traits are passed down:* For more, see C. Darwin, 1868, *The Variation of Animals and Plants Under Domestication,* London: J. Murray.

8 *don't generally blend together and lose identity:* This is technically referred to as particulate inheritance.

8 *Darwin, like virtually every other scientist of his time:* Mayr, 1982, shows that Darwin also had a notion of particulate inheritance, but historically his ideas have been couched primarily in the language of blending inheritance.

8 *the term* gene *became common scientific parlance:* While it is difficult to say exactly when the word *gene* became part of the everyday language of science, the "modern synthesis" of the late 1930s and early 1940s seems reasonable. It was during this period that the fields of evolutionary biol-

ogy, genetics, and systematics had a meeting of the minds. J. Huxley, 1942, *Evolution: The Modern Synthesis,* London: Allen and Unwin.

8 *Genes for this, genes for that:* For more on genetics and human behavior, see D. Hamer and P. Copeland, 1998, *Living with Our Genes: Why They Matter More Than You Think,* New York: Doubleday.

9 *work our way through genetic models of mate choice:* M. Andersson, 1994, *Sexual Selection,* Princeton, N.J.: Princeton University Press.

10 *Darwin's friend George Romanes:* Upon the death of Charles Darwin, Romanes wrote to Darwin's son Francis: "I can derive some consolation from the thought that he died as few men in the history of the world have died—knowing that he had finished a gigantic work, seeing how that work has transformed the thoughts of mankind, and foreseeing that his name must endure to the end of time among the very greatest of the human race." E. Romanes, 1896, *The Life and Letters of George Romanes,* London: Longmans, Green.

10 *There are literally hundreds of definitions of* culture: Kroeber and Kluckhohn review 164 definitions of culture set forth by historians and social scientists. A. L. Kroeber and C. Kluckhohn, 1952, Culture, a critical review of the concepts and definitions, *American Archeology & Ethnology* 47:1–223.

10 *I view culture, and more specifically, cultural transmission:* This definition is similar to those used by Boyd and Richerson in *Culture and Evolutionary Process,* 1985, Chicago: University of Chicago Press: "By 'culture,' we mean the transmission from one generation to the next via teaching and imitation, of knowledge, values and other factors that influence behavior" (p. 2) and "Culture is information capable of affecting individuals' phenotypes which they acquire from other conspecifics by teaching or imitation" (p. 33).

13 *In June 1836, Nathan Rothschild:* This story is relayed in David Landes's 1998 book, *The Wealth and Poverty of Nations,* New York: Norton.

14 *Alan Lill observed that a single male manakin:* A. Lill, 1974, Sexual behavior of the lek-forming white bearded manakin (*Manacus manacus trinitatis* Hartert), *Zeitschrift für Tierpsychologie* 36:1–36.

15 *a single male often gets most of the matings:* J. Höglund and R. Alatalo, 1995, *Leks,* Princeton: Princeton University Press; Andersson, 1994.

15 *Höglund and his colleagues undertook an ingenious experiment:* J. Höglund, R. Alatalo, and A. Lundgren, 1990, Copying the mate choice of others? Observations on female black grouse. *Behaviour* 114:221–236; J. Höglund, R. Alatalo, and A. Lundgren, 1995, Mate-choice copying in black grouse, *Animal Behaviour* 49:1627–1633.

16 *Despite his substantial reputation as a major botanist of his time:* P. Corsi, 1988, *The Age of Lamarck,* Berkeley: University of California Press.

17 *Individuals can cause a change in an organ:* The inheritance of acquired characteristics through use and disuse is often referred to as "soft selection."

18 *Darwin himself accepted Lamarck's ideas on the inheritance of acquired characteristics:* Darwin had problems with many aspects of Lamarck's work, but he felt strongly that the inheritance of acquired characteristics by use and disuse was a real phenomenon (Mayr, 1982).

19 *In* The Nurture Assumption: J. Harris, 1998, *The Nurture Assumption,* New York: The Free Press.

21 *Genes, the argument goes, have deceptively long reaches:* R. Dawkins, 1982, *The Extended Phenotype,* Oxford: Oxford University Press.

22 *thanks to a handful of evolutionary biologists, such a framework:* Here I am referring to people such as Robert Boyd, Peter Richerson, Marcus Feldman, Kevin Laland, Luigi Luca Cavalli-Sforza, E. O. Wilson, and Charles Lumsden.

23 *In their dual inheritance models, theoreticians:* Boyd and Richerson, 1985, p. 4.

24 *Consider the case of a fish less than an inch long:* L. A. Dugatkin, 1996, The interface between culturally-based preferences and genetic preferences: Female mate choice in *Poecilia reticulata, Proceedings of the National Academy of Sciences, U.S.A.* 93:2770–2773.

CHAPTER 2: GENETIC LOVE

27 *No other metaphor in biology has taken hold:* R. Dawkins, 1976, *The Selfish Gene,* Oxford: Oxford University Press.

29 *In his* The Descent of Man and Selection in Relation to Sex: C. Darwin, 1871, *The Descent of Man and Selection in Relation to Sex,* London: J. Murray.

29 *Basically, this difference is due to the fact:* For a good review, see R. Trivers, 1985, *Social Evolution,* Menlo Park, CA: Benjamin Cummings.

32 *Genetic models of female mate choice can be broken down into four groups:* For a concise review of these models see M. Kirkpatrick and M. Ryan, 1991, The evolution of mating preferences and the paradox of the lek, *Nature* 350:33–38.

32 *Evolutionary models don't get simpler than the direct benefit model of female mate choice:* T. Price, D. Schluter, & N. E. Heckman, 1993, Sexual selection when the female benefits directly, *Biological Journal of the Linnean Society* 48:187–211.

32 *In direct benefit models of the evolution of female mate choice:* In principle, of course, the ability to choose males that provide the best resources need not be tied to genes in any real manner, but direct benefit models assume

that it is. For example, in *Behavioral Ecology* (J. Krebs and N. Davies, eds., 1997, Blackwell Science, 4th ed.), the premier graduate text on the subject, Michael Ryan (p. 183) speaks of "mating preference genes" affecting female survival or fecundity.

32 *For example, females may be favored to mate with males:* See Andersson, 1994, chap. 8, for a review of direct benefit models of female mate choice.

33 *Walk around in the woods of southeastern Michigan:* See pp. 367–373 of R. Thornhill and J. Alcock, 1983, *The Evolution of Insect Mating Systems,* Cambridge, Mass.: Harvard University Press. Also see R. Thornhill, 1976, Sexual selection and nuptial feeding behavior in *Bittacus apicalis, American Naturalist* 110:529–548; R. Thornhill, 1980a, Mate choice in *Hylobittacus apicalis* and its relation to some models of female choice, *Evolution* 34:519–538; and R. Thornhill, 1980b, Sexual selection in the black-tipped hangingfly, *Scientific American* 242:162–172.

35 *In other species, however, males wrap up their prey items:* E. L. Kessel, 1955, The mating activities of balloon flies, *Systematic Zoology* 4:97–104; J. M. Cumming, 1994, Sexual selection and the evolution of dance fly mating systems, *Canadian Entomologist* 126:907–920; M. W. Will and S. K. Sakaluk, 1994, Courtship feeding in decorated crickets: Is the spermatophylax a sham? *Animal Behavior* 48:1309–1315.

35 *For the last fifteen of those 2,000 years:* A. P. Møller, 1994, *Sexual Selection and the Barn Swallow,* Oxford: Oxford University Press.

35 *"are almost unique among common European bird species:* Møller, 1994, p. 61.

36 *To determine if females preferred longer-tailed males:* Møller, 1994; A. P. Møller, 1990, Effects of parasitism by the haematophagous mite *Ornithonyssus bursa* on reproduction in the barn swallow *Hirundo rustica, Ecology* 71:2345–2357.

36 *The results of this experimental manipulation:* Females may also be mating with males with good genes, but even so there is a significant direct benefit from low-parasite males.

37 *Good-genes models apply to mating systems in which:* As far back as 1915, R. A. Fisher began to formalize "good-genes" models of female mate choice. R. A. Fisher, 1915, The evolution of sexual preference, *Eugenics Review* 7:184–192.

38 *Amotz Zahavi's "handicap principle":* A. Zahavi, 1997, *The Handicap Principle,* New York Oxford University Press; A. Zahavi, 1975, Mate selection—a selection for a handicap, *Journal of Theoretical Biology* 53:205–214.

39 *For parasite resistance, the argument goes as follows:* This has become known as the Hamilton-Zuk hypothesis, as it was first proposed by William Hamilton and Marlene Zuk. W. D. Hamilton and M. Zuk, 1982, Heritable true fitness and bright birds: A role for parasites, *Science* 218:384–387.

39 *Studies involving dozens of species:* Although some studies fail to provide such support. See Zahavi, 1997.

39 *The handicap principle is far from limited to the case of parasite resistance:* The Hamilton-Zuk hypothesis makes other predictions, which also tend to be supported by many studies.

39 *Godin and I had long worked together on the:* J.-G. Godin and L. A. Dugatkin, 1997, Female mating preference for bold males in the guppy, *Poecilia reticulata, Proceedings of the National Academy of Sciences U.S.A.* 93: 10262–10267.

40 *We examined whether this was the case for guppies:* T. J. Pitcher, D. A. Green, and A. E. Magurran, 1986, Dicing with death: predator inspection behavior in minnow shoals, *Journal of Fish Biology* 28:439–448; L. A. Dugatkin, 1997, *Cooperation Among Animals: An Evolutionary Perspective,* New York: Oxford University Press; L. A. Dugatkin, 1999, *Cheating Monkeys and Citizen Bees: The Nature of Cooperation in Animals and Humans,* New York: The Free Press.

43 *Claus Wedekind, of Bern University:* D. Berreby, 1998, Studies explore love and the sweaty T-shirt, *New York Times,* June 9, p. B14.

43 *These women were then given T-shirts from males:* Women in this study were also given a nasal spray for two weeks prior to the start of the experiment to prevent infection, and they were each given a copy of Patrick Suskind's novel *Perfume* to draw their attention to the importance of odor.

43 *who were not taking oral contraceptives:* Somewhat unexpectedly, Wedekind found that females on oral contraceptives tended to prefer the odors of men with similar MHCs.

43 *In runaway models of animal mate choice, genes in males and genes in females become "linked" to one another:* R. A. Fisher, 1958, *The Genetical Theory of Natural Selection,* New York: Dover; P. O'Donald, 1980, *Genetic Models of Sexual Selection,* Cambridge: Cambridge University Press; Kirkpatrick, 1982, Sexual selection and the evolution of female choice, *Evolution* 36: 1–12.

44 *The best demonstration to date of runaway genetic selection comes from Gerald Wilkinson's work:* G. S. Wilkinson, 1993, Artificial sexual selection alters allometry in the stalk-eyed fly *Cyrtodiopsis dalmanni* (Diptera: Diopsidae), *Genetical Research Camb.* 62:213–222; G. S. Wilkinson and P. Reillo, 1994, Female choice response to artificial selection on an exaggerated male trait in a stalk-eyed fly, *Proceedings of the Royal Society of London* B 255:1–6.

45 *More evidence of runaway sexual selection comes from the guppy:* A. E. Houde, 1992, Sex-linked heritability of a sexually selected character in a natural population of *Poecilia reticulata, Heredity* 69:229-235.

45 *Anne Houde sought to discover a link between the genes that code:* Earlier correlation work suggested such a genetic linkage was at least plausible. A. E. Houde and J. A. Endler, 1990, Correlated evolution of female mating preference and male color pattern in the guppy, *Poecilia reticulata, Science* 248:1405–1408.

46 *At the end of generation four, Houde not only found that males:* A. E. Houde, 1994, Effect of artificial selection on male colour patterns on mating preference of female guppies. *Proceedings of the Royal Society of London* B 256:125–130.

46 *When Felix Breden and Kelly Hornaday ran a similar experiment:* F. Breden and K. Hornaday, 1994, Test of indirect models of selection in the Trinidad guppy, *Heredity* 73:291–297.

46 *dampening the enthusiasm that the guppy system:* There are a myriad of reasons that these studies may have produced differing results. See F. Breden, H. C. Gerhardt, and R. Butline, 1994, Female choice and genetic correlations, *Trends in Ecology and Evolution* 9:343.

46 *The basic premise of the sensory exploitation model is that genes that code:* Also known as "sensory bias" or "sensory drive" models.

46 *Michael Ryan and John Endler, the most vocal proponents of this view:* M. J. Ryan, 1990, Sexual selection, sensory systems and sensory exploitation. *Oxford Surveys in Evolutionary Biology* 7:157–195; M. J. Ryan, 1998, Sexual selection, receiver biases, and the evolution of sex differences, *Science* 281:1999–2003; M. J. Ryan and A. Keddy-Hector, 1992, Directional patterns of female mate choice and the role of sensory biases, *American Naturalist,* 139:s4–s35; J. Endler, 1992, Signals, signal conditions and the direction of evolution, *American Naturalist,* 139:s125–s153; J. Endler and T. McLellan, 1988, The process of evolution: Toward a newer synthesis, *Annual Review of Ecology and Systematics* 19:395–421. Also see the volume 139 supplement to *American Naturalist.* This whole issue is to devoted to sensory bias models.

47 *The sensory exploitation argument goes as follows:* This example follows one illustrated in Kirkpatrick and Ryan, 1991.

47 *One of the most convincing cases of the sensory exploitation:* A. Basolo, 1990, Female preference predates the evolution of the sword in swordfish. *Science* 250:808–811; 1995a, A further examination of a pre-exisiting bias favouring a sword in the genus *Xiphoporus, Animal Behavior* 50:365–375; 1995b, Phylogenetic evidence for the role of a pre-exisiting bias in sexual selection, *Proceedings of the Royal Society of London* 259:307–311.

47 *Female swordtails prefer males with long "swords":* Some phylogenetic work, though controversial, suggests that swords evolved after platyfish and swordtails diverged evolutionarily. A. Meyer, J. M. Morrissey, & M.

Schartl, 1994, Recurrent origin of a sexually selected trait in *Xiphophorus* fishes inferred from a molecular phylogeny, *Nature* 368:539–542; Basolo, 1990; R. Borowsky, M. McClelland, R. Cheng, and J. Welsh, 1995, Arbitrarily primed DNA fingerprinting for phylogenetic reconstruction in vertebrates; the *Xiphophorus* model. *Molecular Biology and Evolution* 12:1022-1032.

47 *Some male swordtails have long swords:* Perhaps because long tails make males appear larger in general. G. Rosenthal and C. Evans, 1998, Female preference for swords in *Xiphophorus helleri* reflects a bias for large apparent size. *Proceedings of the National Academy of Sciences, U.S.A.* 95:4431–4436.

48 *The second example of sensory exploitation comes from two species of frogs:* Physaleaemus pustulosus *and* Physaleaemus coloradorum: M. J. Ryan and A. S. Rand, 1993, Sexual selection and signal evolution: The ghost of biases past, *Philosophical Transactions of the Royal Society of London* 340:187–195.

48 Pustulosus *males, however, are unique in that most add a "chuck" sound:* M. Ryan, 1985, *The Tungara Frog, a Study in Sexual Selection and Communication,* Chicago: University of Chicago Press.

50 *For much of the 1980s, runaway models of female mate choice:* Andersson, 1994.

50 *biologists had to admit that cultural transmission of behavior:* M Kirkpatrick and L. A. Dugatkin, 1994, Sexual selection and the evolutionary effects of mate copying, *Behavioral Ecology and Sociobiology* 34:443–449; K. N. Laland, 1994a, On the evolutionary consequences of sexual imprinting, *Evolution* 48:477–489; K. N. Laland, 1994b, Sexual selection with a culturally transmitted mating preference, *Theoretical Population Biology* 45:1–15. Also see Richerson and Boyd, 1989, for an earlier version of mate-copying models: The role of evolved predispositions in cultural evolution, Or, Human sociobiology meets Pascal's wager, *Ethology and Sociobiology* 10: 195–219.

CHAPTER 3: GUPPY LOVE

54 *I was focused on studying the evolution of cooperation and altruism:* L. A. Dugatkin, 1997, *Cooperation Among Animals: An Evolutionary Perspective,* New York: Oxford University Press; L. A. Dugatkin, 1999, *Cheating Monkeys and Citizen Bees: The Nature of Cooperation in Animals and Humans,* New York: The Free Press.

55 *I was pleasantly surprised to find that guppies:* If you don't count fruit flies.

55 *that there were already dozens of papers on the sexual habits of this species:* To

see just how well studied, see A. E. Houde, 1997, *Sex, Color and Mate Choice in Guppies,* Princeton, N.J.: Princeton University Press.

56 *In 1990, I began the first controlled study of imitation:* L. A. Dugatkin, 1992, Sexual selection and imitation: Females copy the mate choice of others, *American Naturalist* 139:1384–1389.

61 *The protocol for the age/imitation/mate choice experiment:* L. A. Dugatkin and J.-G. Godin, 1993, Female mate copying in the guppy, *Poecilia reticulata:* age dependent effects, *Behavioral Ecology* 4:289–292.

62 *Many studies have listed variables that affect mate choice:* See Houde, 1997.

63 *In 1991, again in collaboration with Jean-Guy Godin:* L. A. Dugatkin and J.-G. Godin, 1992, Reversal of female mate choice by copying in the guppy (*Poecilia reticulata*), *Proceedings of the Royal Society of London* B 249:179–184.

64 *Guppies are not the only fish that display imitation in the context of mate choice:* J. W. A. Grant, and L. D. Green, 1995, Mate copying versus preference for actively courting males by female Japanese medaka (*Oryzias latipes*), *Behavioral Ecology* 7:165–167; K. Witte and M. Ryan, 1998, Male body length influences mate-choice copying in the sailfin molly, *Poecilia latipinna, Behavioral Ecology* 9:534–539.

64 *Somewhat surprisingly, male sailfin mollies also imitate each other's preference for mates:* I. Schlupp and M. Ryan, 1997, Male sailfin mollies (*Poecilia latipinna*) copy the mate choice of other males, *Behavioral Ecology* 8:104–107.

64 *The first formal definition of female mate-choice copying:* S. G. Pruett-Jones, 1992, Independent versus non-independent mate choice: Do females copy each other? *American Naturalist* 140:1000–1009.

65 *individuals need to observe others choose mates for mate-choice copying to have occurred:* Observation here refers not just to vision, but to any sensory input that allows one to know that a mating is taking place.

65 *We must amend Pruett-Jones's definition:* L. A. Dugatkin, 1996, Copying and mate choice, in C. M. Heyes and B. G. Galef, eds., *Social Learning in Animals: The Roots of Culture,* New York: Academic Press.

66 *The more females in a harem, the more opportunities there are for mate copying:* T. H. Clutton-Brock, M. Hiraiwa-Hasegawa, and A. Robertson, 1989, Mate choice on fallow deer leks, *Nature* 340:463–465.

67 *Clutton-Brock and McComb ran two experiments:* T. H. Clutton-Brock and K. McComb, 1993, Experimental tests of copying and mate choice in fallow deer (*Dama dama*), *Behavioral Ecology* 4:191–193; K. McComb and T. H. Clutton-Brock, 1994, Is mate choice copying or aggregation responsible for skewed distributions of females on leks? *Proceedings of the Royal Society of London* B 255:13–19.

68 *A number of studies of female choice in fish clearly show:* For a review of this

phenomenon see L. A. Dugatkin and G. FitzGerald, 1997, Sexual selection, in J.-G. Godin and G. J. FitzGerald, eds., *Behavioral Ecology of Teleost Fishes,* Oxford: Oxford University Press; L. M. Unger and R. C. Sargent, 1988, Alloparental care in the fathead minnow, *Pimephales promelas:* Females prefer males with eggs. *Behavioral Ecology and Sociobiology* 23:27–32.

68 *Initially, people argued that because male sticklebacks, for example:* I. G. Jamieson and P. W. Colgan, 1989, Eggs in the nests of males and their effect on mate choice in the three spined stickleback, *Animal Behaviour* 38:859–865; but see T. Goldschmidt, T. C. Bakker, and E. Feuth-De Bruijn, 1993, Selective copying in mate choice of female sticklebacks, *Animal Behaviour* 45:541-547, for evidence that this is not the case for sticklebacks, *Gasterosteus aculeatus.*

69 *Another explanation for why females choose to mate:* S. Rohwer, 1978, Parental cannibalism of offspring and egg raiding as a courtship strategy, *American Naturalist* 112:429–440.

69 *Suppose, for example, that predatory fish eat a thousand eggs:* In this example, I assume that eggs are distributed randomly and that predators randomly uncover nests.

70 *Steve Shuster and Michael Wade are two excellent evolutionary biologists:* S. M. Shuster and M. J. Wade, 1991, Female copying and sexual selection in a marine isopod crustacean, *Paracerceis sculpta, Animal Behaviour* 42:1071–1078.

70 *Shuster and Wade examined female mate choice in a marine bug:* Males exist in three distinct genetic morphotypes, labeled alpha, beta, and gamma. Alpha males defend breeding territories, beta males resemble breeding females, and small gamma males enter into a male's territory and engage in sperm competition with any males present. S. M. Shuster, 1987, Alternative reproductive behaviors: Three distinct male morphs in *Paracerceis sculpta,* an intertidal isopod from the northern gulf of Mexico, *Journal of Crustacean Biology* 7:318–327; S. M. Shuster, 1989, Female sexual receptivity associated with molting and differences in copulatory behaviour among the three male morphs in *Paracerceis sculpta, Biological Bulletin* 117:331–337.

71 *So until further work is done in this system:* R. Keister, 1979, Conspecifics as cues: A mechanism for habitat selection in the Panamanian grass anole (*Anolis auratus*), *Behavioral Ecology and Sociobiology* 5:323–330.

72 *The quality of a male's territory on a lek then is thought:* See J. Höglund and R. Alatalo, 1995, *Leks,* Princeton, N.J.: Princeton University Press, for more on whether some lek territories are safer than others and whether any territories on leks provide resources.

73 *Robin Gibson and Jack Bradbury have been studying the complex mating dynamics:* See J. W. Bradbury and R. M. Gibson, 1983, Leks and mate choice, in P. Bateson, ed., *Mate Choice,* Cambridge: Cambridge University Press; J. W. Bradbury et al., 1985, Leks and the unanimity of female choice, in P. J. Greenwood, P. H. Harvey, and M. Slatkin, eds., *Essays in Honour of John Maynard Smith,* Cambridge: Cambridge University Press, for early discussions of copying in the sage grouse.

73 *Further, their analysis reveals the snowballing effect:* R. M. Gibson et al., 1991, Mate choice in lekking sage grouse: The roles of vocal display, female site fidelity and copying, *Behavioral Ecology* 2: 165–180. Laboratory studies on sage grouse show that observing another female copulate is essential for copying to occur. M. Spurrier, M. Boyce, and F. Bryan, 1994, Lek behavior in captive sage grouse, *Centrcercus urophasianus, Animal Behaviour* 47:303–310.

74 *There are many species of lekking birds (as well as other animals):* Höglund and Alatalo, 1995. For laboratory evidence of mate copying in other (probably non-lekking) birds, see B. G. Galef and D. J. White, 1998, Mate-choice copying in Japanese quail, *Coturnix cotturnix japonica, Animal Behaviour* 55, 545–552.

75 *as in the great snipe and the pied flycatcher:* P. Fiske, J. A. Kalas, and S. Saether, 1996, Do female great snipe copy each other's mate choice? *Animal Behaviour* 51:1355–1362; T. Slagsvold and H. Viljugrein, 1999, Mate choice copying versus preference for actively courting males by pied flycatchers, *Animal Behaviour* 57, 679–686.

80 *David Buss ran a cross-cultural survey on human mate choice:* D. M. Buss, 1989, Sex differences in human mate preference: Evolutionary hypotheses tested in 37 cultures, *Behavioral Brain Sciences* 12:1–49; D. Buss, 1994, *The Evolution of Desire,* New York: Basic Books.

80 *Song learning in birds varies across species:* P. Slater, L. Eales, and N. Clayton, 1988, Song learning in zebra finches: Progress and prospects, *Advances in the Study of Behavior* 18:1–33; D. Kroodsma and E. Miller, eds., 1982, *Acoustic Communication in Birds,* New York: Academic Press; P. Marler and P. Mundinger, 1971, Vocal learning in birds, in H. Moltz, ed., *Ontogeny of Vertebrate Behavior,* New York: Academic Press; P. Marler and S. Peters, 1982, Long-term storage of stored bird songs prior to production, *Developmental Psychobiology* 15:369–378.

81 *Todd Freeberg undertook a series of fascinating experiments:* T. Freeberg, 1996, Assortative mating in captive cowbirds is predicted by social experience, *Animal Behavior* 52:1129–1142; T. Freeberg, 1998, The cultural transmission of courtship patterns in cowbirds, *Molothrus ater, Animal Behaviour* 56:1063–1073; M. West et al., 1983, Cultural transmission of cowbird

(*Molothrus ater*) song: Measuring its development and outcome, *Journal of Comparative Psychology* 97:327–337; D. Eastzer, A. King, and M. West, 1985, Patterns of courtship between cowbird subspecies: Evidence for positive assortment, *Animal Behaviour* 33:30–39: T. Freeberg, A. King, and M. West, 1995, Social malleability in cowbirds (*Molothrus ater artemisiae*): species and mate recognition in the first two years of life, *Journal of Comparative Psychology,* 109:357–367.

83 *These birds were divided into two categories for separate treatments:* There were two SD/SD/SD groups and two SD/SD/IN groups. During testing, females from SD/SD/SD1 were put with males from SD/SD/SD2 and so on.

84 *Given that there are many species of songbirds:* Another fascinating case of multigenerational transmission of birdsong is found in one species of Darwin's finches (*Geospiza fortis*). In this species, song is passed from grandfather to father to son, and females avoid mating with males that sound like their own fathers. B. R. Grant and P. R. Grant, 1996, Cultural inheritance of song and its role in the evolution of Darwin's finches, *Evolution* 50:2471–2487.

CHAPTER 4: THE MEANING OF CULTURE

88 *The scientific process is fundamentally about generating models of the world:* T. Kuhn, 1962, *The Structure of Scientific Revolutions,* Chicago: University of Chicago Press.

88 *Robert Boyd and Peter Richerson have noted:* A. L. Kroeber and C. Kluckhohn, 1952, Culture, a critical review of the concepts and definitions, *American Archeology & Ethnology* 47, 1–223.

88 *Following Boyd and Richerson, culture:* R. Boyd and P. Richerson, 1985, *Culture and the Evolutionary Process,* Chicago: University of Chicago Press, p. 33. This definition is not all that different from that provided by John Bonner in his book, *The Evolution of Culture in Animals,* Princeton, N.J.: Princeton University Press, 1980. Bonner notes, "By culture I mean the transfer of information by behavioral means, most particularly by the process of teaching and learning."

90 *Approximately 80 percent of the female guppies:* This figure varies from study to study, but is usually in the 75 to 90 percent range.

91 *So it would be great to know as much as possible:* Boyd and Richerson, 1985, write at great length about the interaction of individual and social learning in shaping cultural transmission.

93 *At the most basic level, behavioral ecologists and psychologists:* See the following for reviews: T. D. Johnstone, 1982, The selective costs and benefits of

learning, *Advances in the Study of Behavior* 12:65–106; S. J. Shettleworth, 1984, Learning and behavioral ecology, in J. Krebs and N. Davies, eds., *Behavioural Ecology,* Oxford: Blackwell Scientific; R. Balda, I. Pepperberg, and A. Kamil, eds., 1998, *Animal Cognition in Nature,* San Diego: Academic Press; R. Dukas, ed., 1998, *Cognitive Ecology,* Chicago: University of Chicago Press; S. Shettleworth, 1998, *Cognition, Evolution and Behavior,* New York: Oxford University Press.

94 *That primarily verbal argument has been challenged:* D. Stephens, 1991, Change, regularity and value in the evolution of learning, *Behavioral Ecology* 2:77–89.

94 *Rather than think of environmental predictability as a single force:* Although Stephens, 1991, refers to his own model as one "containing the simplest assumptions that allow a meaningful discussion of learning," it is clear that it is simple only if you know a lot of mathematics. For example, Stephens prefaces his results section as follows: "The problem is now one of studying the long-term dynamics of the 12 difference equations in a situation in which a Markov chain . . ." I do some mathematical modeling, and even I don't think that is simple.

95 *Mind you, this table simplifies many of Stephens's findings:* For example, the table summarizes the extreme cases of Stephens's model quite handily, but does not do justice to intermediate values of between-generation and within-lifetime predictability.

97 *Boyd and Richerson's mathematics can become overwhelming:* To their credit, Boyd and Richerson, 1985, do place their mathematical equations in "boxes" that are set aside from the text, so as to give the reader the option of working through the math or relying on verbal descriptions.

98 *Robert Gibson and Jacob Höglund have suggested two possible reasons:* R. M. Gibson and J. Höglund, 1992, Copying and sexual selection, *Trends in Ecology and Evolution* 7:229–232.

99 *Just as carpenters have many tools in their toolbox:* D. Lendram, 1986, *Modeling in Behavioral Ecology,* Portland, Ore.: Timber Press; M. Mangel and C. Clark, 1988, *Dynamic Modeling in Behavioral Ecology,* Princeton, N.J.: Princeton University Press; J. Krebs, and N. Davies, eds., 1997, *Behavioural Ecology: An Evolutionary Approach,* 4th edition, Sunderland, Mass.: Sinauer Associates; L. A. Dugatkin and H. K. Reeve, eds., 1998, *Game Theory and Animal Behavior,* Oxford: Oxford University Press.

101 *These species are not only full of variance in male mating success:* Bradbury, Vehrencamp, and Gibson found that independent choice of mates on the part of females was unable to account for the huge variance in male reproductive success commonly found in lekking species: J. W. Bradbury, S. L. Vehrencamp, and R. Gibson, 1985, Leks and the unanimity of fe-

male choice. In P. J. Greenwood, P. H. Harvey, and M. Slatkin, eds., *Essays in Honour of John Maynard Smith,* Cambridge: Cambridge University Press.

101 *Wade and Pruett-Jones examined a population:* M. J. Wade and S. G. Pruett-Jones, 1990, Female copying increases the variance in male-mating success, *Proceedings of the National Academy of Sciences, USA* 87:5749–5753.

101 *"Female copying increases the frequency of extreme values:* Ibid., p. 5751.

102 *model is built so that data from natural populations can be handily plugged in:* "When the number of males is large and the sex ratio is approximately unity, there is a simple and sensitive test for the possible existence of female copying. First, testing the observed distribution of matings among males with the expected under random mating (from the binomial distribution) will test for nonrandom mating. Second, calculating the ratio of the observed proportion of males that do not mate with those that mate once estimates s (*the copying parameter*). This estimates the degree of female copying that could account for the observed deviation from random expectation, assuming that female copying is the only factor responsible for nonrandom mating." Wade and Pruett-Jones, 1990, p. 5751. This technique is used in S. M. Shuster and M. J. Wade, 1991, Female copying and sexual selection in a marine isopod crustacean, *Paracerceis sculpta, Animal Behaviour* 42:1071–1078. It is worth noting that this model does not require that females actually observe males mating per se, but rather just that prior male success affects current probability of mating. As such, observation of mating is not necessary for the model to work.

102 *Game theory models of animal culture look at the evolution:* J. Maynard Smith, 1982, *Evolution and the Theory of Games,* Cambridge: Cambridge University Press; Dugatkin and Reeve, 1998.

103 *George Losey and his colleagues built a computer simulation:* G. S. Losey, F. G. Stanton, Jr., T. M. Telecky, W. A. Tyler III, and Zoology 691 Graduate Seminar Class, 1986, Copying others, an evolutionarily stable strategy for mate choice: A model. *American Naturalist* 128:653–664.

103 *Sushil Bikhchandani, David Hirsheifer, and Ivo Welch recently extended the game theory approach:* S. Bikhchandani, D. Hirshleifer, and I. Welch, 1992, A theory of fads, fashion custom and cultural change as information cascades, *Journal of Political Economy* 100:992–1026; D. Hirshleifer, 1995, The blind leading the blind: Social influence, fads, and information cascades, in K. Ieurulli and M. Tommasi, eds., *The New Economics of Human Behaviour,* New York: Cambridge University Press; S. Bikhchandani, D. Hirshleifer, and I. Welch, 1998, Learning from the behavior of others: Conformity, fads, and informational cascades, *Journal of Economic Perspectives* 12:151–170.

104 *Future work in this particular area of modeling may prove very useful:* For more on game theory models of cultural transmission, see C. Findlay, C. Lumsden, and R. Hansell, 1989, Behavioral evolution and biocultural games: Vertical cultural transmission, *Proceedings of the National Academy of Sciences* 86:568–572; C. S. Findlay, R. I. C. Hansell, and C. Lumsden, 1989, Behavioral evolution and biocultural games: Oblique and horizontal cultural transmission, *Journal of Theoretical Biology* 137:245–269; A. Marks, J. C. Deutsch, and T. Clutton-Brock, 1994, Stochastic influences, female copying and the intensity of sexual selection on leks, *Journal of Theoretical Biology* 170:159–162; L. A. Dugatkin and J. Höglund, 1995, Delayed breeding and the evolution of mate copying in lekking species, *Journal of Theoretical Biology* 39:215–218.

104 *Mark Kirkpatrick and I developed population genetic models of mate copying:* M. Kirkpatrick and L. A. Dugatkin, 1994, Sexual selection and the evolutionary effects of mate copying, *Behavioral Ecology and Sociobiology* 34:443–449.

105 *Now that cultural runaway models exist:* For more on population genetic models of cultural transmission, see Boyd and Richerson, 1985; P. Richerson and R. Boyd, 1989, The role of evolved predispositions in cultural evolution or, Human sociobiology meets Pascal's wager, *Ethology and Sociobiology* 10:195–219; K. Aoki, 1989, A sexual-selection model for the evolution of imitative learning of song in polygynous birds, *American Naturalist* 134:599–612; K. Aoki, 1990, A shifting balance type model for the origin of cultural transmission, in N. Takahata and J. F. Crow, eds., *Population Biology of Genes and Molecules,* Tokyo: Baifukan; C. Findlay, 1991, The fundamental theorem of natural selection under gene-culture transmission, *Proceedings of the National Academy of Sciences* 88:4874–4876; K. N. Laland, 1994, On the evolutionary consequences of sexual imprinting, *Evolution* 48:477–489; K. N. Laland, 1994, Sexual selection with a culturally transmitted mating preference, *Theoretical Population Biology* 45:1–15; M. R. Servedio and M. Kirkpatrick, 1996, The evolution of mate choice copying by indirect selection, *American Naturalist* 148:848–867.

107 *If you want children to act in a certain manner:* In *The Nurture Assumption* (New York: The Free Press, 1998), Judith Harris goes as far as arguing that parents contribute to their children via genes, but after that peers, not parents, have the most influence on shaping a child's personality (for good or bad).

107 *These means of transmitting information:* For a thorough treatment of oblique, vertical, and horizontal transmission, see L. L. Cavalli-Sforza

and M. W. Feldman, 1981, *Cultural Transmission and Evolution: A Quantitative Approach,* Princeton, N.J.: Princeton University Press.

109 *Consider this rather odd story about yams:* Boyd and Richerson, 1985, p. 269; W. Bascom, 1948 Ponape prestige economy, *Southwestern Journal of Anthropology* 4:211–221.

111 *"Suppose that at some earlier time Ponapeans did not devote any special effort to growing large yams:* Bascom, 1948.

112 *Even the most basic understanding of the Darwinian process:* D. R. Vining, 1986, Social versus reproductive success: The central theoretical problem of human sociobiology, *Behavioral and Brain Sciences* 9:167–216.

113 *If one assumes that IQ is a measure of intelligence:* R. B. Zajonc, 1976, Family configuration and intelligence, *Science* 192:227–236.

CHAPTER 5: MEME AGAIN

116 *Contenders for this title include Charles Lumsden and E. O. Wilson's* cultragen: C. Lumsden and E. O. Wilson, 1981, *Genes, Mind and Culture,* Cambridge: Harvard University Press; F. Cloak, 1975, Is cultural ethology possible? *Human Ecology* 3:161–182; L. L. Cavalli-Sforza and M. W. Feldman, 1981, *Cultural Transmission and Evolution: A Quantitative Approach,* Princeton, N.J.: Princeton University Press.

116 *The undisputed champion, however, is the concept of the* meme: R. Dawkins, 1976, 1989, *The Selfish Gene,* 1st and 2nd eds., Oxford: Oxford University Press; R. Dawkins, 1982, *The Extended Phenotype,* Oxford: Oxford University Press; D. C. Dennett, 1991, *Consciousness Explained,* Boston: Little, Brown; R. Brodie, 1996, *Virus of the Mind: The New Science of the Meme,* Seattle: Integral Press; A. Lynch, 1996, *Thought Contagion,* New York: Basic Books; R. Dawkins, 1998, *Unweaving the Rainbow,* Boston: Houghton Mifflin; S. Blackmore, 1999, *The Meme Machine,* Oxford: Oxford University Press.

116 *We need a name for the new replicator:* Dawkins, 1976, p. 206.

117 *An element of a culture that may be considered to be passed on by a: Oxford University Dictionary.*

117 *A contagious information pattern that replicates by parasitically infecting human minds: Journal of Memetics* (http://www.cpm.mmu.ac.uk/jom-emit/).

117 *A meme should be regarded as a unit of information:* Dawkins, 1982, p. 109.

117 *A meme is whatever it is that is passed on by imitation:* Blackmore, 1999, p. 43.

117 *A unit of cultural inheritance. Hypothesized as analogous to the particulate gene:* Formally defined in Dawkins, 1982.

118 *With its own on-line journal and popular books: Journal of Memetics*

(http://www.cpm.mmu.ac.ukjom-emit); Brodie, 1996; Lynch, 1996; Blackmore, 1999.

118 *As a sample, here is a section from the memetic lexicon Web site:* http://pespmc1.vub.ac.be/memlex.html.

119 *A replicator has three properties:* Dawkins, 1976.

119 *Dawkins sums up replicators nicely as:* Blackmore, 1999, p. xvi.

122 *As Dawkins notes, "A meme has its own opportunity":* Dawkins, 1982, p. 110.

123 *David Buss has found that both of these phenomena:* D. M. Buss, 1989, Sex differences in human mate preference: Evolutionary hypotheses tested in 37 cultures, *Behavioral and Brain Sciences* 12:1–49; D. Buss, 1994, *The Evolution of Desire,* New York: Basic Books.

124 *Evolutionary psychology explains such behaviors as the product of a:* J. H. Barkow, L. Cosmides, and J Tooby, eds., 1992, *The Adapted Mind: Evolutionary Psychology and the Generation of Culture,* New York: Oxford University Press; D. Buss, 1999, *Evolutionary Psychology,* Boston: Allyn and Bacon.

124 *The most reasonable default assumption is:* L. Cosmides, J. Tooby, and J. Barkow, 1992, Introduction: Evolutionary psychology and conceptual integration, in their *The Adapted Mind: Evolutionary Psychology and the Generation of Culture,* New York: Oxford University Press.

125 *Our brains are set up to handle Pleistocene hunter-gatherer-like conditions:* This is naturally something of an oversimplification of the domain general view, but it is not all that far off from the basic approach adopted.

127 *in the context of a "social contract" or in its absence:* Cosmides, Tooby, and Barkow, 1992.

127 *The Darwinian algorithm approach argues that for scorekeeping problems of equal difficulty:* J. Tooby and L. Cosmides, 1989, Evolutionary psychology and the generation of culture, Part I; Theoretical considerations, *Ethology and Sociobiology* 10:29–49; L. Cosmides and J. Tooby, 1989, Evolutionary psychology and the generation of culture, Part II: Case study: A computational theory of social exchange, *Ethology and Sociobiology* 10:51–98.

129 *"Our Memes is who we are":* Blackmore, 1999, p. 22.

129 *The thesis of this book is that what makes us different:* Ibid., p. 3.

129 *If we define memes as transmitted by imitation:* Ibid., p. 50.

129 *Memetic evolution means that people are different:* Ibid., p. 35.

130 *"Could imitation have been the key to what set our ancestors:* Ibid., p. xii.

130 *Why the strong sense that animals lack memes?* Despite all the quotations from *The Meme Machine* that animals do not possess memes, Blackmore argues that there is one exception to this rule. She believes that memes are present in birdsong, which she argues involves true imitation.

130 *But a small group of influential (and very vocal) psychologists:* C. Heyes, 1996, Introduction: Identifying and defining imitation, in C. M. Heyes and B. G. Galef, eds., *Social Learning in Animals: The Roots of Culture,* New York: Academic Press.

131 *For Blackmore, "Imitation necessarily involves:* Blackmore, 1999, p. 52.

131 *I will also use the term "imitation" in the broad sense:* Ibid., p. 7.

133 *Consider the case that Eberhard Curio and his colleagues examined:* E. Curio, U. Ernest, and W. Vieth, 1978, Cultural transmission of enemy recognition: One function of mobbing, *Science* 202:899–901.

134 *This is especially true if this meme first took hold in older, dominant individuals:* For more on memes and the importance of status, see Blackmore, 1999.

134 *we have a replicator by Blackmore's definition:* Based on her discussion of great tits and milk bottles, I am assuming that Blackmore would still not view "friarbirds are predators" as a meme. The reason is that one could argue that the blackbirds already knew how be to scared; Curio and his colleagues just showed them what to be scared of. From my perspective, however, all that matters is that "friarbirds are predators" appears to meet the criteria that a replicator must meet.

134 *let us return to my long-time lab mate, the guppy:* I could have set up this hypothetical experiment in the streams and tributaries of Trinidad just as easily as in the lab. I simply felt it would be easier to portray in a controlled environment.

135 *Dawkins speculates on the potential power of memes:* Dawkins, 1976, page 206.

CHAPTER 6: ARE YOU MY TYPE?

140 *I ran what might best be called a "titration" experiment:* L. A. Dugatkin, 1996, The interface between culturally-based preferences and genetic preferences: Female mate choice in *Poecilia reticulata, Proceedings of the National Academy of Sciences, U.S.A.* 93:2770–2773.

141 *The results were somewhat surprising:* R. Weiss, 1996, Guppy or yuppie, looks aren't everything, *Washington Post,* April 29, 1996, p. A3.

142 *Sure enough, that made all the difference, and drab males:* L. A. Dugatkin, 1998, Genes, copying and female mate choice: Shifting thresholds, *Behavioral Ecology* 9:323–327.

144 *When given the option, sailfin molly females:* C. Marler and M. Ryan, 1997, Origin and maintenance of a female mating preference, *Evolution* 51:1244–1248. For more on the general nature of female preference for large males, see M. J. Ryan and A. Keddy-Hector, 1992, Directional patterns of female mate choice and the role of sensory biases, *American Naturalist* 139:s4–s35.

145 *Using the basic protocol employed in the guppy experiments:* K. Witte and M. Ryan, 1998, Male body length influences mate-choice copying in the sailfin molly, *Poecilia latipinna, Behavioral Ecology* 9:534–539.

145 *Just as in the guppy titration experiment, culturally obtained information:* I am using phrases like "pushed the female" not in the literal sense, but rather to set a dramatic tone for what is to follow.

146 *In an ingenious experiment, Ingo Schlupp, Cathy Marler, and Michael Ryan:* I. Schlupp, C. Marler, and M. Ryan, 1994, Males benefit by mating with heterospecific females, *Science* 263:373–374.

148 *When he is not doing mathematical models:* H. Whitehead, 1998, Cultural selection and genetic diversity in matrilineal whales, *Science* 82:1708–1711.

150 *As a result of hitchhiking, decreases in cultural variation lead to decreases in mtDNA diversity, and we have our gene-culture interaction:* Whitehead found that this was not due to differences in mutation rate, population size, or generation time.

150 *a case in point being that mtDNA was central to the "Eve" hypothesis:* For example, see L. Vigilant, M. Stoneking, H. Harpending, K. Hawkes, and A. Wilson, 1991, African populations and the evolution of human mitochondrial DNA, *Science* 253:1503–1507.

151 *and their intellectual godfather is Charles Darwin:* J. Weiner, 1995, *The Beak of the Finch: A Story of Evolution in Our Time,* New York: Vintage Books; B. R. Grant and P. Grant, 1989, *Evolutionary Dynamics of a Natural Population,* Chicago: University of Chicago Press; P. R. Grant, 1991, Natural selection and Darwin's finches, *Scientific American* 265: 82–87.

151 *Among the many problems the Grants have tackled is the role:* B. R. Grant and P. R. Grant, 1996, Cultural inheritance of song and its role in the evolution of Darwin's finches, *Evolution* 50:2471–2487.

151 *All species of finch on the Galapagos probably diverged from a common ancestor:* P. R. Grant, 1994, Population variation and hybridization: Comparison of finches from two archipelagos, *Evolutionary Ecology* 8:598–617.

152 *What is surprising is that although they are considered to be different species:* P. R. Grant and B. R. Grant, 1994, Phenotypic and genetic effects of hybridization in Darwin's finches, *Evolution* 48: 297–316.

152 *Not so for the ground finch and the cactus finch:* B. R. Grant and P. R. Grant, 1993, Evolution of Darwin's finches caused by a rare climatic event, *Proceedings of the Royal Society of London* 251: 111–117.

152 *Of 482 females sampled, the vast majority:* Ninety cactus finches and 392 ground finches.

152 *Cultural transmission of song allows females to recognize individuals:* This barrier can be "leaky." See Grant and Grant, 1996, for the implications on species recognition.

153 *In most cases, breeding with relatives (inbreeding) tends to be selected against:* W. M. Shields, 1982, *Philopatry, Inbreeding, and the Evolution of Sex,* Albany: State University of Albany Press; N. W. Thornhill, 1991, An evolutionary analysis of rules regulating human inbreeding and marriage, *Behavioral and Brain Science* 14:247–293; N. W. Thornhill, ed., 1993, *The Natural History of Inbreeding and Outbreeding: Theoretical and Empirical Perspectives,* Chicago: University of Chicago Press.

154 *The concept of "fluctuating asymmetry" is the rage in both evolutionary and psychological work:* For a book-length treatment of this subject, see A. P. Møller and J. P. Swoller, 1997, *Developmental Stability and Evolution,* Oxford: Oxford University Press. For a more concise review, see A. P. Møller and R. Thornhill, 1998, Bilateral symmetry and sexual selection: A meta-analysis, *American Naturalist* 151:174–192.

155 *One way of measuring this would be to subtract:* Usually researchers take the absolute value of the difference between scores, and hence any asymmetry is always reflected in a positive number.

157 *First and foremost, there is evidence that the degree of:* A. P. Møller, 1990, Fluctuating asymmetry in male sexual ornaments may reliably reveal male quality, *Animal Behaviour* 40:1185–1187; R. Thornhill and K. Sauer, 1992, Genetic sire effects in the fighting ability of sons and daughters and mating success of sons in the scorpion fly *(Panorpa vulgaris) Animal Behaviour* 43:255–264; A. P. Møller and R. Thornhill, 1997, A meta-analysis of the heritability of developmental stability, *Journal of Evolutionary Biology* 10:1–16; A. P. Møller and R. Thornhill, 1997, Developmental stability is heritable: Reply, *Journal of Evolutionary Biology* 10:69–76.

157 *For example, when choosing symmetric males:* B. Leung and M. Forbes, 1996, Fluctuating asymmetry in relation to stress and fitness: Effects of trait type as revealed by meta-analysis, *EcoScience* 3:400–413; A. P. Møller, 1996, Parasitism and developmental stability of hosts: A review, *Oikos* 77:189-196; Møller and Swaddle, 1997; R. Thornhill and A. P. Møller, 1997, Developmental stability, disease and medicine, *Biological Reviews* 72:497–548.

158 *Males that end up with territories in the center:* M. Hovi, R. Alatalo, J. Höglund, A. Lundberg, and P. Rintamaki, 1994, Lek centre attracts black grouse female, *Proceedings of the Royal Society of London* 258:303–305.

159 *Males on center territories are indeed more symmetric than males on the edge of an arena:* When measuring the length of tarsal bones. P. Rintamaki, R. Alatalo, J. Höglund, and A. Lundberg, 1997, Fluctuating asymmetry and copulation success in lekking black grouse, *Animal Behaviour* 54:265–269.

159 *Symmetry appears to be a very good cue for a male's genetic quality:* R. Alatalo, J. Höglund, A. Lundberg, P. Rintamaki, and B. Silverin, 1996, Testosterone and male mating success on the black grouse leks, *Proceedings of the Royal Society of London* 263:1697–1702.

161 *In addition, other studies demonstrate that human mate preferences are actually:* D. Buss, 1994, *The Evolution of Desire,* New York: Basic Books; D. M. Buss, 1989, Sex differences in human mate preference: Evolutionary hypotheses tested in 37 cultures, *Behavioral and Brain Sciences* 12:1–49; H. Bernstein, T. Lin, and P. McClellan, 1982, Cross- vs. within-racial judgments of attractiveness, *Perception and Psychophysics* 32:495–503; M. Cunningham, R. Roberts, A. Barber, P. Druen, and C. Wu, 1995, Their ideas of attractiveness are, on the whole, the same as ours: Consistency and variability in the cross-cultural perception of female attractiveness, *Journal of Personality and Social Psychology* 68:261–279.

161 *In a 1998 study, Steve Gangestad and Randy Thornhill:* S. W. Gangestad and R. Thornhill, 1998, Menstrual cycle variation in women's preferences for the scent of symmetrical men, *Proceedings of the Royal Society of London* 265:927–933.

162 *Sure enough, women showed a stronger preference:* This held true only for women who were not taking birth control pills.

162 *Other work has shown that both men and women viewed more symmetric individuals:* K. Grammar and R. Thornhill, 1994, Human facial attraction and sexual selection: The role of symmetry and averageness, *Journal of Comparative Psychology* 108:233–242; T. Shackleford and R. Larsen, 1997, Facial asymmetry as an indicator of psychological, emotional and physiological distress, *Journal of Personality and Social Psychology* 72:456–466. For more examples, see Table I of Møller and Thornhill, 1998. Of course, not all studies suggest that symmetry is attractive: J. Swaddle and I. Cuthill, 1995, Asymmetry and human facial attractiveness: Symmetry may not always be beautiful, *Proceedings of the Royal Society of London* 261:111–116.

162 *The fact that individuals prefer symmetric mates may help explain:* J. A. Simpson et al., 1999, Fluctuating asymmetry, sociosexuality and intrasexual competitive tactics, *Journal of Personality and Social Psychology* 76:159–172.

163 *The direct evidence centers on more symmetric individuals' having:* C. Neugler and M. Ludman, 1996, Fluctuating asymmetry and disorders of developmental origin, *American Journal of Medical Genetics* 66:15–20; R. Thornhill and A. P. Møller, 1997, Developmental stability, disease and medicine, *Biological Reviews* 72:497–548; P. A. Parsons, 1990, Fluctuating asymmetry: An epigenetic measure of stress, *Biological Reviews*

65:131–145; M. Polak and R. L. Trivers, 1994, The science of symmetry in biology, *Trends in Ecology and Evolution* 9:122–124; T. A. Markow and K. Wandler, 1986, Fluctuating dermatographic asymmetry and the genetics of liability to schizophrenia, *Psychiatry Research* 19:325–328; G. Livshits and E. Kobyliansky, 1991, Fluctuating asymmetry as a possible measure of developmental homeostasis in humans: A review, *Human Biology* 63:441–446; N. Peretz, P. Ever-Hadani, P. Casamassimo, E. Eidelman, C. Shelfart, and R. Hagerman, 1988, Crown size asymmetry in males with FRA (X) or Martin-Bell syndrome, *American Journal of Medical Genetics* 30:185–190.

163 *In terms of indirect benefits, symmetric individuals have higher:* F. Furlow et al., 1997, Fluctuating asymmetry and psychometric intelligence, *Proceedings of the Royal Society of London* 264:823–829.

163 *and females report more orgasms with symmetric versus asymmetric sexual partners:* R. Thornhill et al., 1995, Human female orgasm and male fluctuating asymmetry, *Animal Behaviour* 50:1601–1615.

165 *The data are largely concentrated in studies of identical twins:* See C. Reynolds et al., 1996, Models of spouse similarity: Applications to fluid ability measured in twins and their spouses, *Behavior Genetics* 26:73–88, and references therein for more on "assortative mating" and genetics.

166 *This isn't shocking, as new ways of framing critical ideas in a discipline:* T. Kuhn, 1962, *The Structure of Scientific Revolutions,* Chicago: University of Chicago Press.

167 *after individuals observe others in a fight, they behave differently paired with those they have observed:* This sort of thing is far from wild speculation. Observation has been demonstrated to play a role in animal aggression; see L. A. Dugatkin, 1997, Winner effects, loser effects and the structure of dominance hierarchies, *Behavioral Ecology* 8:583–587; L. A. Dugatkin, 1998, Breaking up fights between others: A model of intervention behaviour, *Proceedings of the Royal Society of London* 265:443–437; L. A. Dugatkin, 1998, A model of coalition formation in animals, *Proceedings of the Royal Society of London* 265:2121–2125.

CHAPTER 7: ANIMAL CIVILIZATION

170 *The gorilla case above is completely fictitious and comes straight from Crichton's novel* Congo: M. Crichton, 1995, *Congo,* New York: Mass Market Paperbacks.

171 *and learned to wash their own sweet potatoes at early ages:* S. Kawamura, 1959, The process of sub-culture propagation among Japanese macaques, *Primates,* 2:43–60. This work is summarized in T. Nishida, 1987, Local

traditions and cultural transmission, in B. Smuts, D. Cheney, R. Seyfarth, R. Wrangham, & T. Struhsaker, eds., *Primate Societies,* Chicago: University of Chicago Press.

172 *Michael Huffmann, for example, has found another case:* M. Huffman, 1996, Acquisition of innovative cultural behaviors in nonhuman primates: A case study of stone handling, a socially transmitted behavior in Japanese macaques, in: C. M. Heyes and B. G. Galef, eds., *Social Learning in Animals: The Roots of Culture,* London: Academic Press.

172 *But stone use, and the cultural transmission of stone use, are not uniquely human:* M. Huffman and D. Quiatt, 1986, Stone handling by Japanese macaques: Implications for tool use of stone, *Primates* 27:427–437.

173 *The reason for opening a discussion of nonmating examples of cultural:* In fact, despite seven different long-term studies of culture in chimpanzees, very little in the way of controlled work on cultural evolution has been done in this group in their natural habitats. A. Whiten, J. Goodall, W. McGrew, T. Nishida, Y. Sugiyama, C. Tutin, R. Wrangham, and C. Boesch, 1999, Cultures in chimpanzees, *Nature* 399:8682–8685. For a list of other examples of primate imitation see A. Whiten and R. Ham, 1992, On the nature and evolution of imitation in the animal kingdom: Reappraisal of a century of research, *Advanced Study in Behavior* 21:239–263.

173 *In fact, culture has been shown to be a powerful force in:* Some of the examples I draw on in this chapter can be found in more technical form in two excellent academic books on social learning, Heyes and Galef, 1996; T. R. Zentall and B. G. Galef, 1988, *Social Learning: Psychological and Biological Perspectives,* Hillsdale, NJ: Erlbaum.

174 *to determine what they should and shouldn't add to their diets:* See table 7.1 of LeFebrve and Palameta for a list of all social learning and foraging experiments through 1988: L. LeFebrve, and B. Palameta, 1988, Mechanisms, ecology and population diffusion of socially learned, food-finding behavior in feral pigeons, in Zentall and Galef, 1988.

175 *Rat fetuses can actually sense what type of food their mother is eating:* B. Galef and M. Clark, 1972, Mother's milk and adult presence: two factors determining initial dietary selection by weaning rats, *Journal of Comparative Physiology and Psychology,* 78:220–225; B. Galef and P. Henderson, 1972, Mother's milk: A determinant of feeding preferences of weaning rat pups, *Journal of Comparative Physiology and Psychology,* 78:213–219; B. Galef and D. Sherry, 1973, Mother's milk: A medium of transmission of cues reflecting the flavor of mother's diet, *Journal of Comparative Physiology and Psychology,* 83:374–378.

175 *It's worth noting that the food used in the experiment demonstrating this ability*

was cloves of garlic: P. Hepper, 1988, Adaptive fetal learning: Prenatal exposure to garlic affects postnatal preferences, *Animal Behaviour* 36:935–936.

175 *The story of social learning and food preferences in rats began:* P. Ward and A. Zahavi, 1973, The importance of certain assemblages of birds as "information centres" for finding food, *Ibis* 115:517–534.

176 *Galef and his colleague Stephen Wigmore tested this hypothesis:* B. Galef and S. Wigmore, 1983, Transfer of information concerning distant foods: A laboratory investigation of the "information-centre" hypothesis, *Animal Behaviour* 31:748–758.

177 *Train a rat to interact for long periods:* B. G. Galef, 1989, Enduring social enhancement of rats' preferences for the palatable and the piquant, *Appetite* 13:81–92.

178 *Galef and his colleagues injected some unlucky rats with lithium chloride:* B. Galef et al., 1983, A failure to find socially mediated taste aversion learning in Norway rats (*R. norvegicus*), *Journal of Comparative Psychology* 97:358–363.

178 *A similar (but not nearly as detailed) story can be told of culture and foraging in the rat's dreaded enemy, the cat:* W. Wyrwicka, 1978, Imitation of mother's inappropriate food preference in weaning kittens, *Pavlovian Journal of Biological Science* 13:55–72.

178 *Why social learning is so finely tuned in helping rats choose what to eat:* Galef, for example, has suggested that natural selection has simply favored learning about foods, but that selective forces on learning to avoid poisons may have been weaker, producing no socially mediated avoidance to new toxins. B. G. Galef, 1996, Social enhancement of food preferences in Norway rats: A brief review, in Heyes and Galef, 1996.

179 *J. Fisher and Robert Hinde put forth the notion that this new behavior:* J. Fisher and R. Hinde, 1949, The opening of milk bottles by birds, *British Birds* 42:347–357; R. Hinde and J. Fisher, 1951, Further observations on the opening of milk bottles by birds, *British Birds* 44:393–396.

179 *The mischief seemed to be breaking out simultaneously:* R. Hinde, 1982, *Ethology: Its Nature and Relations with Other Sciences,* Oxford: Oxford University Press, p. 108.

180 *David Sherry and Galef (of imitation and rats fame):* D. Sherry and B. G. Galef, 1984, Cultural transmission without imitation: Milk bottle opening by birds, *Animal Behavior* 32:937–939; D. Sherry and B. G. Galef, 1990, Social learning without imitation: More about milk bottle opening by birds, *Animal Behaviour* 40:987–989.

181 *When Palameta and LeFebrve did their first studies on copying and foraging in*

pigeons: B. Palameta and L. LeFebrve, 1985, The social transmission of a food-finding technique in pigeons: What is learned? *Animal Behaviour* 33:892–896.

182 *There is an interesting twist to the pigeon story. In many group-living animals:* C. J. Barnard, ed., 1984, *Producers and Scroungers,* London: Croom Helm/ Chapman Hall.

182 *Layered on to the imitation and foraging story we have seen in pigeons:* L. A. Giraldeau and L. LeFebrve, 1986, Exchangeable producer and scrounger roles in a captive flock of feral pigeons: A case for the skill pool effect, *Animal Behaviour* 34:797–803.

183 *This apparent paradox caused Giraldeau and LeFebrve to examine:* L. A. Giraldeau and L. LeFebrve, 1987, Scrounging prevents cultural transmission of food-finding in pigeons, *Animal Behaviour* 35:387–394.

184 *Consider this statement by Paul Rozin:* P. Rozin, 1988, Social learning about food by humans, in Zentall and Galef, 1988.

185 *Back in 1938, K. Duncker examined the food preferences of preschoolers:* K. Duncker, 1938, Experimental modification of children's food preferences through social suggestion, *Journal of Abnormal and Social Psychology* 33:489–507.

186 *In a similar but more controlled vein, L. Birch:* L. Birch, 1980, Effects of peer models' food choices and eating behaviors in preschoolers' food preferences, *Child Development* 51:14–18.

186 *such as chili pepper:* P. Rozin and D. Schiller, 1980, The nature and acquisition of a preference for chili peppers by humans, *Motivation and Emotion* 4:77–101.

186 *The answer probably lies in the fact that the foods that we eat:* J. Billing and P. W. Sherman, 1998, Antimicrobial functions of spices: Why some like it hot, *Quarterly Review of Biology* 73:3–49. Billing and Sherman argue that spices serve an important function in terms of protecting us from various microbial diseases. Further, after surveying 4,578 recipes from around the world, Billing and Sherman demonstrate that people that live in areas with the greatest probability of microbial diseases (e.g., hot areas where food spoilage is a problem) have the greatest intake of antimicrobial spices.

187 *Consider the 1993 field experiment undertaken by E. Hanna and A. Meltzoff:* E. Hanna and A. Meltzoff, 1993, Peer imitation by toddlers in laboratory, home and day-care contexts: Implications for social learning and memory, *Developmental Psychology* 29:701–710.

187 *Other infants "sat around a table, drinking juice, sucking their thumbs":* A. Meltzoff, 1996, The human infant as imitative generalist: A 20-year

progress report on infant imitation with implications for comparative psychology, in Heyes and Galef, 1996.

188 *Enter Kevin Laland, Kerry Williams, and the guppy:* K. N. Laland and K. Williams, 1997, Shoaling generates social learning of foraging information in guppies, *Animal Behaviour* 53:1161–1169. For an experiment with similar protocol (but using rat foraging as the system of choice), see B. G. Galef, Jr., and C. Allen, 1995, A new model system for studying behavioural traditions in animals, *Animal Behaviour* 50:705–717.

190 *Back in 1984, Gene Helfman and Eric Schultz ran an experiment:* G. Helfman and E. Schultz, 1984, Social transmission of behavioral traditions in a coral reef fish, *Animal Behaviour* 32:379–384.

192 *Eberhard Curio and his colleagues did just that:* E. Curio et al., 1978, Cultural transmission of enemy recognition: One function of mobbing, *Science* 202:899–901.

192 *Blackbirds, like many other bird species, undertake a fascinating antipredator:* S. A. Altmann, 1956, Avian mobbing behavior and predator recognition, *Condor* 58:241–253; T. A. Sordahl, 1990, The risks of avian mobbing and distraction behavior: An anecdotal review, *Wilson Bulletin* 102:349–352.

193 *In other words, how long is the "cultural transmission chain" in blackbirds?* For example, if a naive bird became a teacher and then its new subject went on to become a teacher, our chain length would be two.

194 *In controlled laboratory experiments with juvenile rhesus monkeys:* S. Mineka, M. Davidson, M. Cook, and R. Keir, 1984, Observational conditioning of snake fear in rhesus monkeys, *Journal of Abnormal Psychology* 93:355–372.

194 *Interestingly, it made no difference whether the individuals they:* M. Cook, S. Mineka, B. Wolkenstein, and K. Laitsch, 1985, Observational conditioning of snake fear in unrelated rhesus monkeys, *Journal of Abnormal Psychology* 94:591–610.

195 *Tim Caro and Marc Hauser tackle the question of animal teaching:* T. M. Caro and M. D. Hauser, 1992, Is there teaching in nonhuman animals? *Quarterly Review of Biology* 67:151–174.

201 *Arriving at a general theory of cultural transmission, and its implications for understanding behavior, is just a matter of time:* L. A. Dugatkin, 1999, *Cheating Monkeys and Citizen Bees: The Nature of Cooperation in Animals and Humans,* New York: The Free Press.

Index

INDEX

INDEX

INDEX

INDEX

INDEX

INDEX

INDEX